Java Card for E-Payment Applications

For quite a long time, computer security was a rather narrow field of study that was populated mainly by theoretical computer scientists, electrical engineers, and applied mathematicians. With the proliferation of open systems in general, and of the Internet and the World Wide Web (WWW) in particular, this situation has changed fundamentally. Today, computer and network practitioners are equally interested in computer security, since they require technologies and solutions that can be used to secure applications related to electronic commerce. Against this background, the field of computer security has become very broad and includes many topics of interest. The aim of this series is to publish state-of-the-art, high standard technical books on topics related to computer security. Further information about the series can be found on the WWW at the following URL:

> http://www.esecurity.ch/serieseditor.html

Also, if you'd like to contribute to the series by writing a book about a topic related to computer security, feel free to contact either the Commissioning Editor or the Series Editor at Artech House.

Recent Titles in the Artech House
Computer Security Series

Rolf Oppliger, Series Editor

Computer Forensics and Privacy, Michael A. Caloyannides

Demystifying the IPsec Puzzle, Sheila Frankel

Electronic Payment Systems for E-Commerce, Second Edition, Donal O'Mahony, Michael Pierce, and Hitesh Tewari

Information Hiding Techniques for Steganography and Digital Watermarking, Stefan Katzenbeisser and Fabien A. P. Petitcolas, editors

Internet and Intranet Security, Second Edition, Rolf Oppliger

Java Card for E-Payment Applications, Vesna Hassler, et al.

Non-repudiation in Electronic Commerce, Jianying Zhou

Secure Messaging with PGP and S/MIME, Rolf Oppliger

Security Fundamentals for E-Commerce, Vesna Hassler

Security Technologies for the World Wide Web, Rolf Oppliger

Software Fault Tolerance Techniques and Implementation, Laura L. Pullum

For a complete listing of the *Artech House Computing Library,*
turn to the back of this book.

Java Card for E-Payment Applications

Vesna Hassler
Martin Manninger
Mikhail Gordeev
Christoph Müller

Pedrick Moore
Technical Editor

AH

Artech House
Boston • London
www.artechhouse.com

Library of Congress Cataloging-in-Publication Data
Java card for e-payment applications / Vesna Hassler ... [et al.].
        p.   cm. — (Artech House computer security series)
    Includes bibliographical references and index.
    ISBN 1-58053-291-8 (alk. paper)
    1. Java (Computer program language).   2. Smart cards.
I. Hassler, Vesna.
QA76.73.J38 J3638   2001
005.13'3—dc21

                                                        2001045735

British Library Cataloguing in Publication Data
Java card for e-payment applications. — (Artech House computer security series)
    1. Java (Computer program language)   2. Smart cards
    3. Electronic commerce—Security measures
I. Hassler, Vesna   II. Moore, Pedrick
005.7'1262

    ISBN 1-58053-291-8

Cover design by Igor Valdman

International Standard Book Number: 1-58053-291-8
Library of Congress Catalog Card Number: 2001045735

10 9 8 7 6 5 4 3 2 1

Contents

Preface

Smart cards—also referred to as microprocessor cards—are a well-known technology that has been in use in Europe and some parts of Asia for almost two decades. Existing applications range from building access control to mobile phones (Subscriber Identity Module cards). Note that smart card technology encompasses all applications of the smart card microprocessor, even if it is not embedded on a plastic card, but within a USB token instead. Because microprocessor technology is constantly improving, the number of potential applications for smart cards is growing. Moreover, for some application areas, such as those with high security requirements, smart cards represent state-of-the-art security tokens.

For some time, the development of smart card–based services was impeded by the lack of a broadly accepted programming language for card applications on the one hand, and the lack of a common framework for using card services from the application terminal on the other. The Java Card specification by Sun Microsystems now offers a "dialect" of a popular programming language that is used across many different platforms. Meanwhile, two interindustry initiatives, PC/Smart Card and OpenCard Framework (OCF), provide frameworks for integrating smart cards with the desktop. This book addresses both frameworks but places special emphasis on Java Card and the OCF because they provide a platform-independent environment for using multiple-application smart cards.

This is not the first book about smart cards, but it is different from other books we have seen. Part I gives a general introduction to the cards, covering their logical structure, security issues, and relevant standards. Part II introduces Java Card as a platform for developing on-card applications in a specially tailored subset of Java, with a special emphasis on security issues. Part III deals with smart card APIs and explains the basics of OCF programming. Finally, Part IV delivers a practical example using Java Card and the OCF to develop an on-card application, an OCF card service, and a terminal application using the card service. The application is based on a real-world credit/debit example inspired by the Europay/MasterCard/Visa specification, which is currently the most important specification for credit/debit card payment applications worldwide.

The book is composed in such a way that it can be used, for example,

- By IT managers and CTOs to who would like an overview of the technology;
- By engineers as a quick introduction to smart cards and especially Java Card and OCF;
- As a textbook for a university course.
- Some names used throughout the book are trademarks. We do not provide information about the trademarks, so please check at the corresponding registration office if necessary.

We hope that you will enjoy reading the book and learning about smart cards. We would appreciate any feedback. You can reach us at jcbook@yahoogroups.com.

Vesna Hassler
Martin Manninger
Mikhail Gordeev
Christoph Müller
Vienna/Perm
November 2001

Acknowledgments

We are deeply grateful to all those who supported us in writing this book. Here we mention just some of them (in alphabetical order): A-SIT, Secure Information Technology Center–Austria; Oliver Fodor, Vienna, Austria; Gemplus, France; Giesecke & Devrient, Germany; Ruth Harris, Artech House, United Kingdom; Institute for Applied Information Technology and Communications (IAIK), Graz University of Technology, Graz, Austria; Tim Pitts, Artech House, United Kingdom; Siemens AG Austria, Vienna, Austria; and XSoft GmbH, Vienna, Austria.

We are especially grateful to Giesecke & Devrient for providing us with their Sm@rtCafé Java Card toolkit, which we used to develop the on-card application presented in the book. Special thanks also to the IAIK for giving us their Java Cryptography Extension (IAIK-JCE 2.61, see http://jcewww.iaik.at) free of charge to develop the application for the book.

Part I
Smart Cards

Smart cards became a widely used technology during the 1990s. They are very common in the European Union and parts of Asia, where they are employed for banking transactions, mobile phones, access control, and more. Up to now, however, the educational systems for engineers in the fields of electrical engineering and information technology have rarely included courses on smart card technology. Part I of this book is intended to explain the basics of smart cards as well as the most important standards in this area. Special emphasis is given to Java Cards as they provide the most important hardware-independent multiapplication platform.

1

Smart Card Basics

The term *smart card* is commonly used for a plastic card containing a piece of golden contact plate (see example in Figure 1.1), although this usage is not strictly correct from a technical point of view. The more general term for cards containing integrated circuits (ICs) is integrated circuit cards (ICCs) or, in common speech, also chip cards. ICCs that are tamper-resistant can serve as security tokens in many different fields of application. They communicate with card readers or card terminals,[1] which also supply the necessary power and clock to the card. In principle, two independent building blocks determine the function of an integrated circuit card: the communication interface and the internal logic.

1.1 Logic of Integrated Circuit Cards

In terms of their logic, one should differentiate between *memory cards* and *microprocessor cards*. Figure 1.2 shows the internal structure of a memory card, which consists mainly of some hardwired communication and security logic and a read/write memory, usually an electrically erasable programmable read-only memory (EEPROM). Typical functions of the security logic of a memory card are optional write protection of certain memory areas and a primitive authentication mechanism, such as allowing access to the memory only after a password has been entered. This is a major advantage over

---

1. See Chapter 6 for explanations on card readers and card terminals.

3

Figure 1.1 Example of an integrated circuit card. (*Source:* Austria Card.)

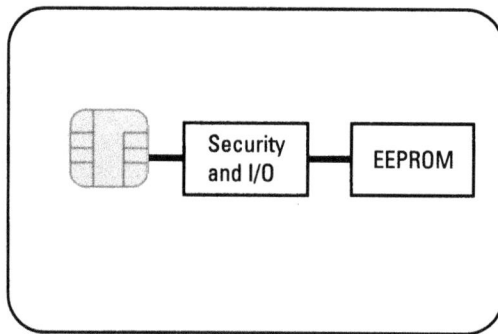

Figure 1.2 Internal structure of a memory card.

magnetic-strip cards, where the magnetic-strip data are always readable and hence can easily be copied. Memory cards were the first ICCs deployed in significant numbers. Until now they have served as disposable prepaid telephone cards, as access cards, and as authentication instruments in various other applications with low or medium security levels.

Microprocessor cards contain (as shown in Figure 1.3) a full microcomputer consisting of a central processing unit (CPU), read-only memory (ROM), an EEPROM, random access memory (RAM), and an input/output peripheral unit that handles communication with the outside world. Because all of these elements are combined in one integrated circuit, this specific type of IC is also called a *microcontroller*. As of 2001, typical technical standards

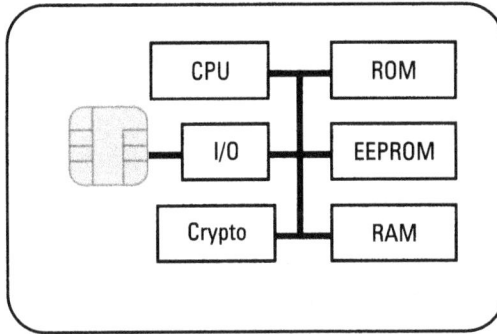

Figure 1.3 Internal structure of a smart card.

are an 8-bit or 16-bit CPU, 16 to 64 Kbytes of ROM, 4 to 64 Kbytes of EEPROM, and 256 bytes to 1 Kbyte of RAM. Microprocessor cards have the ability to carry out instructions fetched from ROM or EEPROM and do all of the traditional data processing that a microcomputer is capable of. Thus microprocessor cards are the ones that really deserve the name "smart" card. They contain a layered software architecture where applications are defined on top of a card operating system.

Today we often differentiate between smart cards with and without cryptography coprocessors. Such units are needed to perform the algorithms of asymmetric cryptography, especially the RSA algorithm,[2] within a reasonable time. The CPU of a smart card operates at much lower clock frequencies than do today's desktop computers and cannot achieve sufficient performance for complex calculations. The latest high-end smart cards, however, work with a 32-bit architecture and internal clock generation instead of using the (slower) external clock to overcome this limitation.

Smart cards serve as banking cards (including electronic purses), as mobile phone cards (the so-called *Subscriber Identity Modules* (SIM), usually found in the smaller plug-in format), as digital signature cards, and as authentication instruments in various other applications with medium or high security levels. In 2000, more memory cards were still in use than microprocessor cards, but microprocessor cards show higher growth rates and are estimated to take the lead in the next few years.

---

2. RSA, named after its inventors Rivest, Shamir, and Adleman, is the most widely used asymmetric cryptographic algorithm. See also Chapter 2.

1.2   Communication Interface of Integrated Circuit Cards

From the previous section we know that the presence of a contact plate says nothing about the processing capabilities of a card. Likewise, the absence of contact plates does not indicate that the card is a "dumb" piece of plastic. In addition to the ordinary communication via galvanic contacts, which is explained in detail in Chapter 5, a contactless communication interface is being used more and more. In this case the power and clock are provided, and the data transferred over a certain distance, via inductive (or sometimes capacitive) coupling between a card terminal and a card. This means that a card need not be inserted into a terminal but merely be held close enough to establish communication. Figure 1.4 shows an example of a contactless card, in which the chip and the connected coil ("antenna") can clearly be seen.

Contactless ICCs and contactless card readers are more able to withstand mechanical and electrical impact, which makes them well suited for outdoor applications. Furthermore, they are perfect for ticketing applications such as that required for public transport and for controlling building access and electronic "punch" clocks because of their ability to communicate over a

Figure 1.4 Contactless ICC. (*Source:* Philips Semiconductors.)

distance, which reduces the total transaction time for the user. Several standards have been defined for contactless cards, starting with the rarely used *Close-Coupling Cards* standard, ISO 10536. The most widespread contactless cards as of 2002 are the *proximity cards* according to ISO 14443 with a typical range of 8 cm. Yet systems operating over a distance of 80 cm and more are available and may be useful in those applications where the user should not need to express his will by consciously moving the card near a terminal. The latest such standard is named *Vicinity Cards* (ISO 15693).

Besides contact-only and contactless cards, different types of combinations are also possible. First, one can simply put two independent chips into one card, each with its own communication interface. This is called a hybrid contact and contactless card and requires less development effort, but lacks internal communication between the two chips. This solution is often chosen if approved components have to be combined, for example, a contactless memory chip for building-access control or ticketing applications and a contact microprocessor chip providing a personal computer sign-on function. In such a case, none of the terminals and background systems that have already been developed for the particular applications need be altered in the slightest detail.

Second, special *dual-interface chips* equipped with a contact and a contactless communication interface have been developed. They offer many more possibilities, but require the development of special card operating systems. A good example is a combined card with a contactless electronic ticket for public transport, and a payment function, such as a debit or credit card or an electronic purse. With a hybrid card solution, the ticket-vending machine would be responsible first for performing the payment transaction via the contacts and then for storing the ticket via the contactless interface. This means, of course, that the ticket-vending machine would need both communication interfaces, but there is an additional requirement. From the cardholder's point of view, a dual-interface card solution is preferred because it can, with a single card command, perform both the payment transaction and the ticket loading in one step. Thus any opportunity for cheating ticket-vending machines is eliminated.

1.3 Smart Card Operating Systems

Like that of a desktop computer, a smart card operating system has to be specific to the underlying hardware features. Its tasks are file management, communication, and execution of commands. A card operating system has to meet special requirements resulting from the limited resources available on the smart card IC. In particular, it must fit into the ROM and execute commands quickly despite mediocre CPU computing power. Consequently, mechanisms like dynamic memory allocation are usually not implemented, and large parts of today's operating systems are still written in assembly languages instead of higher level languages.

Because the operating system resides in the ROM, it must be absolutely error free. Any critical error would require the replacement of all cards issued and the destruction of all ICs in that particular production run. To avoid this, many operating systems allow patches for their major parts to be loaded into the EEPROM if certain security conditions are fulfilled. Nonetheless, smart card operating systems have to be developed and tested much more carefully than conventional operating systems.

Most of today's smart card operating systems do not support multiple, independent applications. Any piece of code can access all data and hence different applications could compromise each other. This does no harm as long as the card and all of its applications are under the control of one issuer, but in the future more true multiple-application cards will be demanded. The three most important systems are Java Card, MULTOS, and Windows for Smart Cards.

True multiple-application operating systems implemented on ICs that provide sufficient hardware security to guarantee separated memory areas may even allow a new application code to be loaded after the card has been issued. All three systems provide this option. Java cards accept Java programs compiled into Java byte code, MULTOS cards accept Java or C programs compiled in the Multos Execution Language (MEL), and Windows for Smart Cards cards accept a sort of Visual Basic–type program. MULTOS applications must be registered and certified by a central certification authority, which is a well-structured security architecture but results in additional bureaucracy and fees. Windows for Smart Cards was developed later than the other two systems, but it is expected to become widespread through Microsoft's strategy to support card-based sign-on in all of its newer PC operating systems. The first cards intended to be used with Windows 2000, however, are not equipped with Windows for Smart Cards but with conventional card operating systems.

1.4   Smart Card Life Cycle

The first steps in the life of a smart card (shown in Figure 1.5) are taken by the semiconductor supplier who develops a microcontroller and suitable

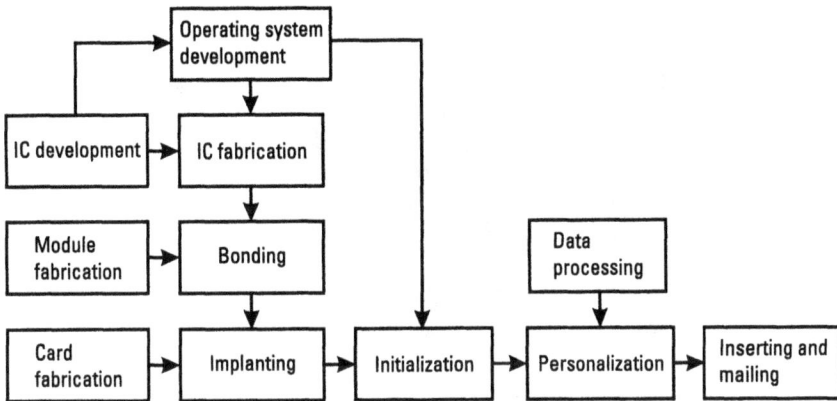

Figure 1.5 Smart card production steps.

manufacturing processes. On top of this hardware platform, the operating system developer can implement all desired functions of the card operating system. Usually the developer will use simulators and emulators during the development phase because of the long manufacturing time of real ICs.

When all emulator tests are positive, the operating system developer will submit the ROM code to the semiconductor supplier, who will then produce the first batch of ICs, incorporate them into packages, and connect them to the contact plate (*bonding*) if appropriate. The bonding of ICs to chip modules may also be done by the card manufacturer. If not, then the chip modules will arrive at the card manufacturer's site on standardized tapes (see Figure 1.6) that can be processed by the chip-module implanting machines.

The cards themselves are made from plastic, particularly from poly-vinylchloride (PVC), polycarbonate (PC), polyethylene-terephthalate (PET), or acrylonitrile-butadiene-styrene (ABS).

High-quality cards require several manufacturing steps, such as printing the designs, laminating a number of plastic layers together, and punching out the standardized card size. Then a cavity is milled into the card where the IC module can be implanted and glued in.

During the initialization process, which is also called the prepersonalization process, the EEPROM parts are written in to complete the operating system. In addition, the file structures and other data not individual to the card are loaded into the card in this step, because initialization machines can usually handle cards faster than personalization machines can.

Once the necessary user data have been gathered and processed by the issuer, an order is sent to the card manufacturer. The personalization

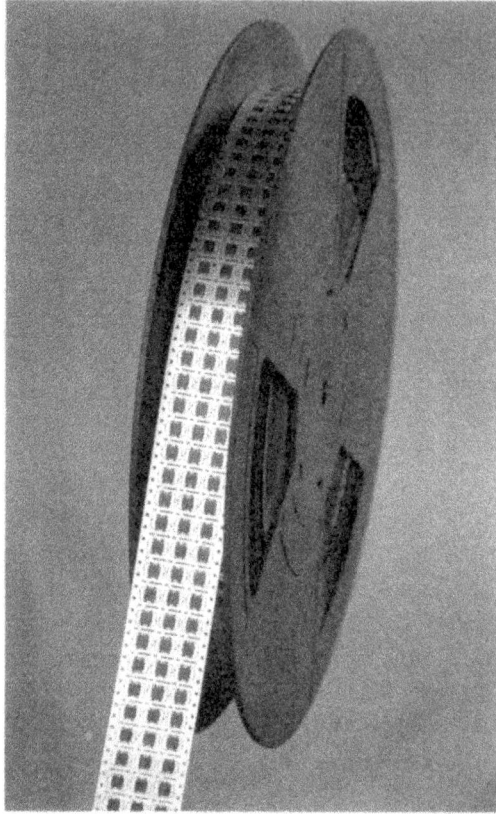

Figure 1.6 ICs on tape. (*Source:* Philips Semiconductors.)

machine writes these specific data into the EEPROM or, by use of a laser or other printing mechanism, onto the surface of the card. The last step concerns shipping, with all its preparations. For example, the card can be glued to a letter containing personal text modules, and supplementary details can be added before everything is put into an envelope and mailed to the final customer.

Once cards are "out in the field" in ordinary use, they can no longer normally be accessed by the issuer. If, however, they are inserted into a terminal connected to the issuer's host, the issuer can update certain parameters. As one of many countermeasures to fraud, system developers tend to implement irreversible deactivation functions, which are then activated by the terminal or by the card itself, for example, on a certain date.

1.5 Integrated Circuit Card Standards

Table 1.1 gives an overview of the most important international standards for integrated circuit cards. Much of their content will be explained in later chapters.

Table 1.1
Integrated Circuit Card Standards

Standard	Date	Title
ISO 7810	1995	Identification Cards—Physical Characteristics
ISO 7816-1	1998	Identification Cards—Integrated Circuit(s) Cards with Contacts—Part 1: Physical Characteristics
ISO 7816-2	1999	Identification Cards—Integrated Circuit(s) Cards with Contacts—Part 2: Dimensions and Locations of Contacts
ISO 7816-3	1997	Identification Cards—Integrated Circuit(s) Cards with Contacts—Part 3: Electronic Signals and Transmission Protocols
ISO 7816-4	1995	Identification Cards—Integrated Circuit(s) Cards with Contacts—Part 4: Interindustry Commands for Interchange
ISO 7816-5	1994	Identification Cards—Integrated Circuit(s) Cards with Contacts—Part 5: Numbering System and Registration Procedure for Application Identifiers
ISO 7816-6	1996	Identification Cards—Integrated Circuit(s) Cards with Contacts—Part 6: Interindustry Data Elements
ISO 7816-7	1999	Identification Cards—Integrated Circuit(s) Cards with Contacts—Part 7: Interindustry Commands for Structured Card Query Language (SCQL)
ISO 7816-8	1999	Identification Cards—Integrated Circuit(s) Cards with Contacts—Part 8: Security Related Interindustry Commands
ISO 7816-9	2000	Identification Cards—Integrated Circuit(s) Cards with Contacts—Part 9: Additional Interindustry Commands and Security Attributes
ISO 7816-10	1999	Identification Cards—Integrated Circuit(s) Cards with Contacts—Part 10: Electronic Signals and Answer to Reset for Synchronous Cards
ISO 10373	1998	Test Methods, Parts 1, 2, 5
ISO 10536	1996	Identification Cards—Contactless Integrated Circuit(s) Cards, Parts 1–3
ISO 14443	2000	Identification Cards—Contactless Integrated Circuit(s) Cards—Proximity Cards, Parts 1–4
ISO 15693	2000	Identification Cards—Contactless Integrated Circuit(s) Cards—Vicinity Cards, Parts 1–2

2

Security Issues

This chapter is dedicated to security, because this is the primary reason why smart cards are used at all. Before dealing with specific questions on the security of smart cards, we shall explain the basic elements of security and cryptography. A reader familiar with these concepts can skip Sections 2.1, 2.2, and 2.3.

Safety and security both express the ability of a system to maintain its intended functionality under external influences. The term *safety* is used for resistance to random events, such as technical breakdown or failure, and other effects resulting from environmental conditions. *Security* refers to resistance to intentional attacks such as espionage or sabotage. Obviously, a secure system is to a certain extent also a safe system, because if intentional actions are prevented, the same actions will also be prevented from happening accidentally.

We now concentrate on security that is harder to achieve, and specifically on systems in the field of information technology. Five characteristics of data or data processing systems can be threatened by attackers:

1. *Availability*. The attacker intends to make the data or the system unusable. Making the system unusable is also called a denial-of-service attack.

2. *Integrity*. The attacker intends to modify data in such a way that the modification will not be noticed.

3. *Confidentiality.* The attacker intends to obtain knowledge of confidential information.

4. *Authenticity/originality.* The attacker intends to copy data and use it as if he were the owner. He does not necessarily need to be able to interpret the data.

5. *Nonrepudiation.* The attacker intends to deny that he has created certain data or sent a certain message.

Many of these threats can be prevented by *cryptography,* which originally meant only "secret writing" but today is known as the science of securing data. The counterpart to cryptography is *cryptanalysis,* which is the science of breaking cryptographic mechanisms. Steganography is another possibility for keeping data secret. In this case, the data are hidden among larger amounts of irrelevant data in such a way that a potential attacker does not even recognize the existence of the secret data. To give an example, the individual bits of an ASCII text message with a total size of 100 bytes can be coded as least significant bits of a 3-Mbyte MPEG-3 file, which can in turn be made available for downloading together with dozens of similar files.

2.1 Symmetric Cryptography

Figure 2.1 shows the basic principle of cryptography. A plaintext message M is encrypted ("enciphered") by use of a mathematical transformation and additional information named key K_E to a ciphertext C that contains all the original information but cannot be interpreted without being decrypted ("deciphered") by use of the reverse mathematical transformation and a corresponding key K_D. The result of the decryption is the original plaintext M.

The keys used in this procedure are of particular importance. More than 100 years ago, Kerckhoff recognized that the strength of a cryptosystem must be based on its keys. It is rarely possible to keep the algorithms of the encryption and decryption transformations confidential. Keys are easier to

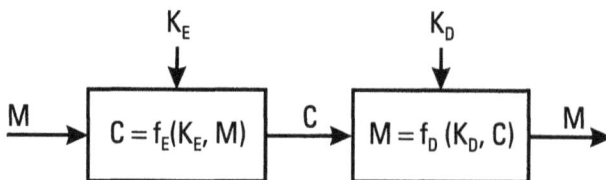

$$K_E \qquad\qquad K_D$$

$$M \longrightarrow \boxed{C = f_E(K_E, M)} \xrightarrow{\ C\ } \boxed{M = f_D(K_D, C)} \xrightarrow{\ M\ }$$

Figure 2.1 Encryption and decryption.

keep confidential, and they can easily be updated (changed) once an attack is suspected.

In symmetric cryptography, the same key $K = K_E = K_D$ is used for encryption and decryption.[1] To transmit a confidential message from point A to point B via an insecure network, both communication partners have to agree on a cryptosystem and a secret key K. Symmetric cryptography is also called secret key cryptography. Of course, the key must not be transmitted via the insecure network but via a secure connection. Transmission of the key need not be synchronous with transmission of the actual communication that is secured with the key.

Two possible actions can help an attacker determine the key. Using cryptanalysis, he can perform a brute force attack that tries all possible key values. If an appropriate key has been chosen, this sort of attack will take too long to be useful. But cryptanalysis provides more intelligent methods than a brute force attack. Some examples are the *known-plaintext attack,* where the attacker knows at least parts of a certain plaintext and the corresponding ciphertext, and the *chosen-plaintext attack,* where the attacker is able to have any plaintext encrypted to its corresponding ciphertext. In these cases, the secret key can be found more easily.

The second possibility for an attacker to determine the key has nothing to do with cryptanalysis but simply with ordinary espionage and theft. Keys that have been found out are called compromised keys. Apart from ordinary (physical) espionage, electronic espionage occurs. Today, most data are stored in computers with insecure operating systems. On these platforms, it is not possible to prevent viruses, Trojan horses, and unwanted parts of application software or the operating system itself from executing. Hence there is also a high risk that key files may be copied. The best solution to this problem is to store keys only in tamper-proof devices with certified software, namely, in smart cards.

Two types of symmetric cryptographic algorithms exist:

- *Stream algorithms* can work on a data stream and encrypt and decrypt individual bits or bytes of the data stream.

- *Block algorithms* encrypt and decrypt data blocks of a certain size; however, block algorithms can also be used in a mode that results in a stream cipher.

---

1. It may not be the case that $K = K_E = K_D$ directly. However, to meet the criteria of a symmetric cryptosystem, K_E and K_D must be (efficiently) computable from one another.

Table 2.1 gives some examples of symmetric algorithms.

Symmetric cryptography is used not only to secure confidentiality; it can also help provide authenticity as well. A message authentication code (MAC) is a cryptographic checksum calculated over a data block. There are several ways to calculate these MACs. If the data block is short enough, it can simply be encrypted and sent to the recipient in both plaintext and cipher-text—but this makes a known-plaintext attack and some other attacks possible. A better and more suitable choice for larger amounts of data is to use a one-way hash function to calculate a short hash value of the data and then to encrypt this value.

Symmetric cryptography has certain advantages regarding mathematical complexity and execution time. Its major disadvantage is the necessity for every two potential communication partners to exchange a key and to keep the key secret. Furthermore, if a key has been compromised, it is difficult to find out which of the communication partners has been the target of the espionage.

Table 2.1
Symmetric Cryptography Algorithms

Algorithm	Block/Stream	Key Length (bits)	Comment
Data Encryption Standard (DES)	Block	56	Standardized in 1977 with FIPS PUB 46; in 2000, still most widely used; at a key length of 56 bits no longer highly secure, but Triple DES with an effective key length of 112 bits provides very good security; common in smart cards
International Data Encryption Algorithm (IDEA)	Block	128	Very good security and faster than Triple DES; not yet widespread because of patents held by ASCOM and consequent license fees
Advanced Encryption Standard (AES)	Block	128, 192, 256	Intended as the successor to DES; standardization planned in 2001; can be implemented on smart cards with small ROM and RAM size
Ron's Code 4 (RC4)	Stream	Variable	Developed by Ron Rivest for RSA Data Security; although not free of license fees, widely used, for example, with Secure Sockets Layer technology

2.2 Asymmetric Cryptography

Asymmetric cryptography uses different keys for encryption and decryption. The encryption key and the decryption key must be related to each other, but it is not possible to calculate one from the other. Thus one of those keys can be made public (public key K_p) and the other kept secret (private key or secret key K_s). Asymmetric cryptography is therefore also known as public key cryptography.

For data encryption, the intended recipient first generates the private/public key pair and sends the public key to an open directory or directly to potential communication partners. Everyone can then use this public key to encrypt data, but not to decrypt them. Decryption can only be done with the corresponding private key. The major advantage of asymmetric cryptography is that the secret key need not be transmitted; it remains at the recipient's key depository, which can be a smart card.

The principle of the private and the public key can be symbolized as a key and a lock (see Figure 2.2). Person *A* wants to receive secured (encrypted) messages, so she produces a number of equal padlocks that can be distributed among her potential communication partners (the public key that can be copied), but she produces only one matching key (the secret key), which she keeps secure. Communication partner *B* can put the message in a chest and secure it with the padlock (encrypt it). On receipt of the chest (encrypted message), person *A* can unlock (decrypt) it with the key.

Besides its use for encryption, asymmetric cryptography is also the basis for digital signatures. A digital signature to a message is calculated with the author's secret key. Assuming that everyone has access to the corresponding public key, the originality of the message can be verified, and with a suitable system concept, nonrepudiation can be achieved as well. Again, it is suitable to sign a hash value of the message to save processing time.

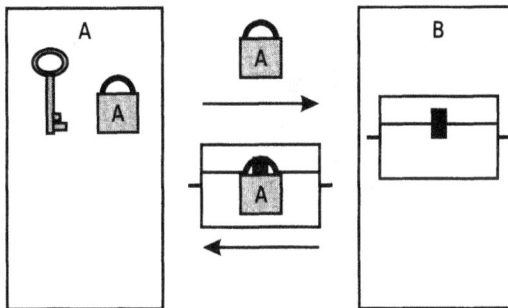

Figure 2.2 Principle of asymmetric cryptography.

Note that digital signatures and hand-written signatures differ more than their common name would imply:

- A hand-written signature is related to one natural person. It is physically related to the signed document, but looks the same on every document, that is, it does not depend on the document content. The document content is generally understandable for the signer without additional equipment.
- A digital signature can be related to a natural person or a legal entity. In any case, it can be produced by anyone and everyone who has access to the secret key. The signature will not vary according to who produced it. Furthermore, a digital signature depends on the content of the document in such a way that even adding a single semicolon to the document causes most of the bytes of the signature to change their values. One disadvantage of digital signatures is the fact that the signer needs technical equipment to be able to understand the document content. For commercial applications, the signer has to trust this viewer equipment and assume that there is no other way to interpret the string of bytes he is going to sign.

Although the need for key exchange via a secure channel has been eliminated, one problem is still unsolved. An attacker M could generate his own key pair K_{SM} and K_{PM} and give his public key K_{PM} to the sender instead of the recipient's key. Figure 2.3 shows the principle of this "man-in-the-middle attack" employing the key and lock symbols introduced above. The padlock sent from person A to person B is intercepted by M and replaced by one of M's own padlocks. Person B believes he has received A's padlock and uses it to secure the message. M intercepts the chest, opens it, and reads the message,

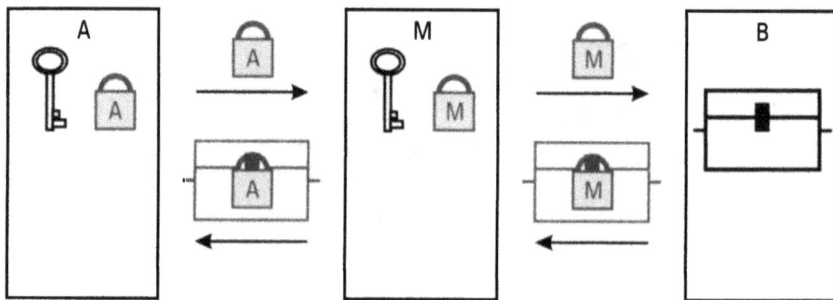

Figure 2.3 Principle of man-in-the-middle attack.

then puts A's original padlock back on the chest and sends the chest to A. Neither person A nor B will find out that the communication is insecure.

Figure 2.4 shows the man-in-the-middle attack in a more formal way. First M intercepts the public key sent from A to B and replaces K_{PA} with K_{PM}. Communication partner B cannot verify that the public key he receives actually belongs to M instead of A. If B uses K_{PM} to encrypt a message, M will intercept it and decrypt the ciphertext using K_{SM}. To remain undiscovered, M can encrypt the plaintext afterward using K_{PA} and send the new ciphertext to A. There is no way for A to know that the message has been read by someone else because it arrives encrypted with the valid K_{PA} and can be decrypted with K_{SA}.

To prevent man-in-the-middle attacks, the relationship of a public key to the owner of the corresponding private key must be secured. This can be done either by submitting the public key via an authentic communication channel (which may be impractical) or by involving a trusted third party (TTP). A centralized TTP is also called a trust center or certification authority, because it certifies the relationship of a public key to its owner. The scheme illustrated in Figure 2.5 involves a notary public (TTP T) who has once visited B and left him a method to verify the authenticity of the notary's signature. When A produces her identical padlocks and her matching key,

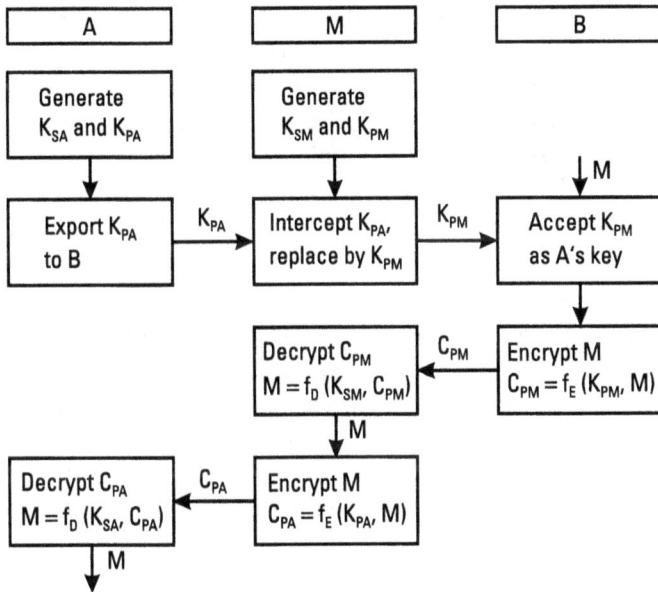

Figure 2.4 Man-in-the-middle attack.

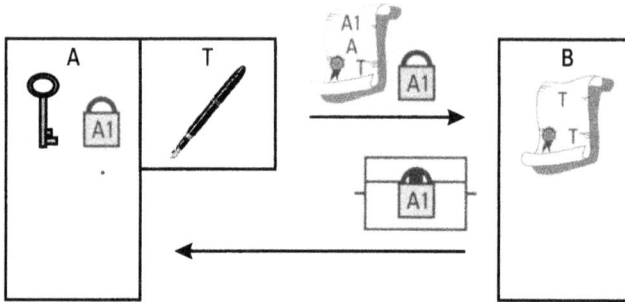

Figure 2.5 Principle of asymmetric cryptography with certificate.

the notary takes one of the padlocks and engraves his signature beside the serial number. The notary issues a certificate saying that the padlock with this serial number was presented to him by *A*. By reading this certificate and comparing the serial number of the padlock, *B* can then verify that this really is *A*'s lock.

Figure 2.6 shows the formal scheme of digital signatures with certificates. Signer *A* generates a private/public key pair and submits the public key K_{PA} via a secure channel to the certification authority, who in turn issues a certificate saying that this key really belongs to *A*. Certificate *Z* is signed with the private key of the certification authority K_{ST} and made public via an open directory service. Then *A* signs a message with K_{SA} and sends it to recipient *B*.

Person *B* retrieves *A*'s certificate from the directory service and verifies its authenticity by use of K_{PT}. The certification authority's public key K_{PT} has to be known by every potential communication partner. Note that, although a secure channel is needed to transmit K_{PT}, no secure channel is necessary between the two communication partners. In the end *B* can be sure of having the original K_{PA} and thus will trust the signed message if verification with K_{PA} gives a positive result.

With the addition of time stamps, digital signatures can also provide nonrepudiation. However, the time stamp must be generated by a TTP. Self-generated time stamps or no time stamps at all do not prevent the signer from claiming that his key was already compromised at the time the digital signature was added to the document.

Besides this centralized approach of setting up a certification authority, decentralized approaches may also be suitable. One example is Pretty Good Privacy's (PGP) *Web of Trust*, where every participant may sign the public keys of others. If *A* signs the public key of *B*, *B* the key of *C*, and *C* the key of *D*, then *E* can trust the public key of *D* even if *E* has only received the trusted

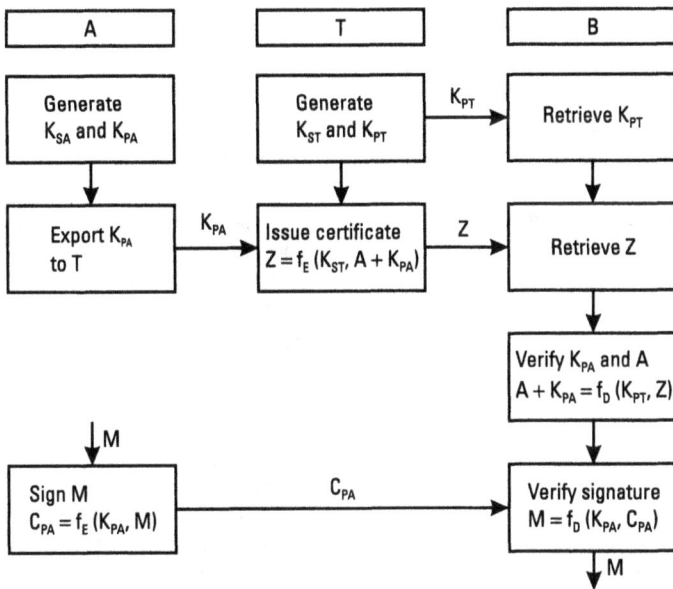

Figure 2.6 Digital signature with certificate.

public key of *A*. The more digital signatures are attached to a PGP public key, the more likely it is to be trusted by potential communication partners without further action. Of course, this network of trust may be corrupted by people who certify incorrect keys. PGP's internal rules require two "half-trusted" signatures to accept a new public key as trusted. Once a key has been compromised, it must be revoked immediately.

Table 2.2 gives some examples of asymmetric algorithms. All of them can use different key lengths. As of 2001, key lengths of 512 bits to 2,048 bits are common.

Elliptic curves can also be used as a basis for asymmetric cryptography. DSA, for example, can be implemented in elliptic curves. Many experts today declare that elliptic curves make shorter key lengths and hence faster processing possible while keeping the same security level. If this assumption is true, the smart cards of the future will not necessarily need expensive cryptography coprocessors to be capable of asymmetric cryptography.[2]

---

2. Another development exists that will lead to smart cards being capable of asymmetric cryptography without having a specialized coprocessor: Within a few years the performance of the card's main CPUs will simply be good enough to calculate 1,024-bit, or even 2,048-bit, RSA encryption and decryption within a reasonable time.

Table 2.2
Asymmetric Cryptography Algorithms

Algorithm	Encryption/ Signature	Key Length (bits)	Comment
Rivest, Shamir, Adleman (RSA)	Encryption/ Signature	1024–2048	Developed 1978 and most widely used algorithm in 2001; no longer patented since September 2000; still secure if key lengths of 1,024 bits or more are applied; common in smart cards
Digital Signature (DSA)	Signature	Variable	Standardized 1994 with FIPS PUB 186; slower than RSA; limited security owing to the maximum key size; several patents exist; rarely used
ElGamal	Encryption/ Signature	Variable	A little slower than RSA and equally secure; not patented but rarely used

The major shortcoming of asymmetric cryptography is still that the performance is much slower than with symmetric cryptography. To overcome this disadvantage, hybrid symmetric/asymmetric encryption schemes have been constructed. One of the communication partners generates a symmetric session key and encrypts it with his asymmetric private key. The recipient can decrypt it using the sender's public key and eventually a related certificate. Afterward both can use this session key to secure their communication with fast symmetric cryptography. If there is only one large message to be encrypted, the session key encrypted with the private key can be attached directly to the symmetrically encrypted message itself.

2.3 Authentication

Authentication is the process of verifying the identity or the group membership that a person (or a device) claims to possess. Figure 2.7 shows the three basic authentication methods. Biometrics, like face or fingerprint comparison, are the most practical methods to use between human beings, but they are complex to implement in machines and usually have a certain error rate. Authentication through possession of an item, such as a cryptographic key, is easier to implement. The key may be passed on, but it should be constructed in a way that makes duplication nearly impossible. The third

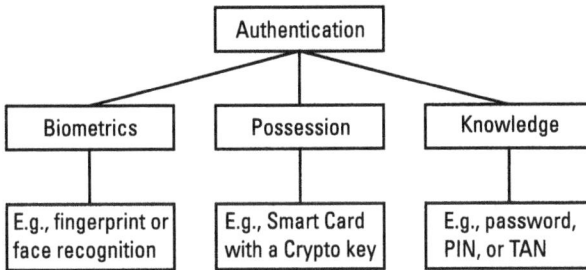

Figure 2.7 Authentication methods.

method, authentication through knowledge, is most common in computer systems. It can be used for authentication of a user to a machine and also between machines.

Typical authentication by stating a user name and a password is not suitable in insecure or untrusted environments, because the secret information needs to be transmitted and thus can be intercepted and duplicated. The same is valid for a personal identification number (PIN), which is nothing but a password that is defined as being known by only one person. One-time passwords such as transaction authentication numbers (TANs) are much better, although their disadvantage is that TAN lists have to be stored somewhere and have to be updated occasionally. Cryptography can be employed to generate authentication tokens dynamically. This is called the *challenge/response authentication method*, in which, for example, the one who has to authenticate receives a random challenge and encrypts it using a secret key. The ciphertext is then transmitted back and can be verified by the communication partner, who holds the same secret key (in the case of symmetric cryptography).

Challenge/response authentication offers the advantage that secret information need not be transmitted during authentication. Interception of such authentication messages is normally of no use for the attacker because the random challenge and its related response are always different. Only if the attacker can intercept the responses to a large part of all possible challenges will she likely be able to use one of them. To prevent this kind of replay attack, challenges are constructed not only from random numbers but contain continuously increasing counter values or time stamps as well.

If both communication partners need to authenticate each other, this can be done by two challenge/response authentications, one after the other. It is more elegant and more secure to combine these steps into one *mutual authentication*. Its message flow is shown in Figure 2.8. First, communication

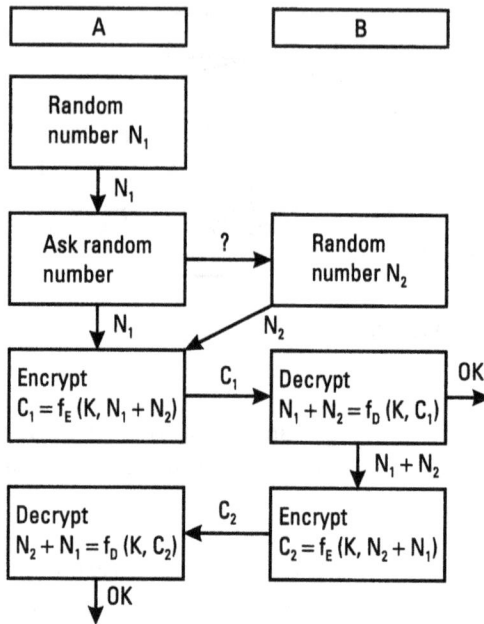

Figure 2.8 Mutual authentication.

partner A generates a random number N_1 and asks communication partner B for a random number N_2. Both random numbers are concatenated to one data package, which is then encrypted with the key K. A transmits the ciphertext C_1 to B, who can decrypt it and verify that the number N_2 is correct. Then B constructs a data package containing both random numbers in reverse order to prevent replay attacks. Communication partner B transmits the encrypted data package C_2 to A, who can in turn decrypt it and verify that the number N_1 is correct. In this way, both communication partners have authenticated each other.

Of course, a challenge/response authentication can be performed with asymmetric cryptography, too. The principles remain the same. The additional benefit is that the secret key exists only once and does not have to be distributed.

In summary, the three levels of knowledge-based authentication methods are as follows:

1. Passwords and similar, where the secret information is known to the sender and the recipient, and is sent over the communication channel between them;

2. Symmetric challenge/response authentication, where the secret information is known to the sender and the recipient; and

3. Asymmetric challenge/response authentication, where the secret information is known only to the sender.

After authenticating a user, a system can grant him certain rights. This process is called *authorization*. Flexible authorization methods are role based, which means that the rights are related to roles, and the users become associated with one or more roles after authentication.

2.4 Smart Card Security

Because the industry claims smart cards to be tamper-resistant or even tamper-proof hardware, considerable effort has been spent on reaching the maximum security level. The use of appropriate cryptography as explained above is just one important element. Cryptography may be used not only on the system level but also for securing internal procedures of the chip. For example, data transport from the EEPROM to the CPU via the internal buses can be encrypted.

Besides cryptography, other software measures are vital. The chip operating system and the chip application software must not contain any backdoors that would allow unauthorized access to secret data. If the operating system allows the loading and execution of new application programs, it must ensure that these programs cannot affect the data or program code of other applications. The operating system should also check the integrity of the relevant memory areas, for example, by calculating checksums of ROM and EEPROM contents and checking the RAM with write and read tests during the chip's startup procedure. A general rule for smart card applications is to allow no more access than necessary. State machines are implemented that define what conditions have to be fulfilled to allow the execution of a command or access to a file.

A smart card IC manufacturer's greatest problem is to ensure that secret information such as the cryptographic keys remains secret. Even the most severe software measures would be in vain if there were any possibility that data could be read from the chip by direct access, bypassing the chip software. Because conventional ICs can be read, for example, with an electron microscope, special hardware security measures are built into smart card ICs. These hardware measures can be divided into passive and active measures.

Passive hardware security measures include shields, layered architecture, and scrambling. It is not too difficult for an attacker to disassemble the chip module and obtain access to the surface of the IC. Shields are additional metallic films covering the semiconductor surface. If an attacker wants to access the semiconductor, he needs to damage the shield, and this will cause the IC to cease working. The architecture of a smart card IC makes critical areas harder to access because they are placed in the deepest levels of the silicon. For example, several other layers have to be penetrated before the ROM is reached. Furthermore, unlike in conventional ROM, the memory cells are produced in such a way that the data content cannot be seen with microscopes. Other hardware measures aim at confusing potential attackers by scrambling. In particular, the memory addresses as well as the order and paths of the internal buses can be scrambled.

Active hardware security measures allow the IC to detect attacks by using various sensors and to react, for example, by putting itself out of operation. The validity of supply voltage and clock frequency in particular are checked, because invalid values can cause malfunctions of the IC that can in turn enable an attacker to bypass other security mechanisms or to gather critical data directly. Another active measure is to check the integrity of the "passivation layer." This is a layer covering the entire semiconductor to protect it from oxidation. Its electrical resistance or capacity can be measured, which helps to detect any modifications. As soon as sensors indicate suspicious conditions, the IC can cease operation, switch to a special error state, or even delete all critical data. Although these sensors allow clever countermeasures, they work only as long as the chip is supplied with power. If someone tries to attack the chip without setting it into operation, it must be protected by the passive security measures.

2.5 Known Attacks on Smart Cards

Although smart card IC developers employ the best security technology available, time and again new attacks are found that can successfully be applied. The three most important attacks developed in recent years are presented below.

In 1996, Ross Anderson and Markus Kuhn showed that they needed nothing more than a sharp knife and a little fuming nitric acid and acetone to expose the semiconductor surface of a smart card IC. To penetrate the passivation layer, microprobing needles can be employed. A focused ion beam workstation costs several million U.S. dollars, but it is capable of severing existing connections inside the IC, establishing new connections, and changing the semiconductor doping. An attacker who understands the internal structure of a certain IC can disconnect the entire CPU except for the program counter and have it count all possible EEPROM addresses. Then he needs only to find the data bus in order to read all of the EEPROM content. Of course, in up-to-date IC architecture, this is not as easy as it may sound (see the passive security measures described above).

Also in 1996, three researchers at Bellcore, Dan Boneh, Richard A. DeMillo, and Richard J. Lipton, published a theoretical method for cracking cryptographic algorithms implemented in hardware. Their work was based on the fact that hardware performing calculations will often produce incorrect results if environmental conditions are causing stress. Comparing these faulty results of cryptographic operations to the correct results yields sufficient information to calculate secret keys with reasonable effort. This "Bellcore attack" was first directed at asymmetric cryptography using algebraic operations. An RSA implemented with the *Chinese remainder theorem* had been found to be extraordinarily vulnerable. Soon Eli Biham and Adi Shamir published a related attack on secret key cryptography like DES. They called this the differential fault analysis (DFA).

Despite the theoretical strength of these attacks, an easy-to-implement countermeasure exists. The IC simply has to check the correctness of its cryptographic calculations before making any result available to the outside.

The latest major attack method was published by Paul Kocher in 1998. The simple power analysis (SPA) uses a variation of the IC's power consumption, especially during the performance of a cryptographic algorithm. In the case of an improperly designed piece of hardware or operating system, the attacker can gain secret information directly from the power consumption graph. SPA-proof smart cards are possible, however, and up-to-date cards fulfill this requirement. The advanced version of this attack is the differential power analysis (DPA). By means of statistical analysis of many power consumption measurements, even well-designed smart cards can be attacked. Although DPA cannot be fully prevented as of 2001, a level of DPA resistance can be achieved that makes these attacks impractical.

Timing attacks are an earlier work of Paul Kocher. He showed that, if the duration of cryptographic operations depends on the input data, then it is possible to find out, for example, RSA secret keys, by measuring and evaluating execution times.

2.6 System Security

We have seen that while smart cards offer the highest level of security, a small risk of successful attacks does remain. It is doubtful whether something like a perfectly tamper-proof device is possible at all. Besides making resistance to all imaginable attacks higher and higher by always employing the latest generation of smart cards ICs and operating systems, it is very important to consider the system environment of a smart card. For example, no one would take the trouble to modify one prepaid phone card to avoid paying the price of telephone calls if the expense for conducting this attack were a million U.S. dollars. On the other hand, if a successful attack led to the possibility of producing and using an unlimited number of faked credit cards, someone would possibly be willing to invest amounts well in excess of a million dollars.

This leads us to the system view, where risk distribution is one of the most important goals. Two requirements have to be met:

1. The information contained in one smart card cannot be sufficient for an attacker to compromise the whole system.

2. The maximum gain resulting from a successful attack on one card shall be less than the expense of this attack.

Similar considerations have to be made for other system components as well. Card terminals in particular are often the targets of attacks (see Chapter 6).

Because, in a well-designed system, the cryptographic keys are the only critical data, several key management concepts have been developed and are commonly employed. First of all, key generation must take place in a secure environment.

In the case of asymmetric cryptography, where the private key never needs to be exported, the best way is to have the card's key pair generated inside the card, which does not even offer a command to export the private key. This method also fulfills the second rule, which says that each card shall have individual keys. To ensure interoperability with all other system components, there must be a set of master keys, but these are only stored in a few high-security components. By use of a master key and an individual attribute of a chip, such as a hardware serial number, a derived key is calculated. This derived key is then used for communication solely with this particular card. Of course, several nested key derivations are possible as well. In addition to derived keys, there may be different versions of every key in the system. If an attack on a master key is suspected, this master key and all of its derived keys can be declared invalid. Instead of having to stop operation for a major security repair, which could entail the exchange of all cards issued, the system can go on with normal operation using the next version of the master key.

3

Security Evaluation Criteria

There is no such thing as a perfectly secure computing or communication system, because it is impossible to prove formally that an arbitrary system is secure. Such proof can be performed only for very simple systems or protocols, and even for those it is very computing intensive. We can, however, verify that a program is correct, which means that it is also possible to build verifiably correct secure systems. This can be achieved by integrating the verification of a system into its specification, design, and implementation. And this is the principle around which security evaluation criteria are built.

The *Department of Defense Trusted Computer System Evaluation Criteria* (TCSEC [1]) was the first collection of security evaluation criteria applied to rate the security offered by a computer system [2]. The TCSEC was developed in the early 1980s. Since the TCSEC and its supporting documents all have differently colored covers, they are often referred to as the *Rainbow Series,* and the TCSEC itself as the *Orange Book.* The evaluation criteria developed in other countries built on the concepts of the TCSEC but tried to make the criteria more flexible and thus better adaptable to the rapidly evolving information technology [3]. The TCSEC defines seven sets of evaluation criteria called classes (D, C1, C2, B1, B2, B3, and A1), grouped into four divisions (D, C, B, and A). Each criteria class covers four aspects of evaluation: security policy, accountability, assurance, and documentation. The criteria for these four areas become more detailed from class to class, and form a hierarchy whereby D is the lowest (minimal protection) and A1 the highest (verified design).

The *Information Technology Security Evaluation Criteria* (ITSEC [4]), a joint development project of France, Germany, The Netherlands, and the United Kingdom, was published by the European Commission in 1991. The ITSEC and TCSEC have many similar requirements, but also some important differences. The ITSEC places more emphasis on integrity and availability, and attempts to provide a uniform approach to the evaluation of both products and systems. In addition, the ITSEC differentiates between the effectiveness of security-enforcing functions and mechanisms and their correctness [2]. The ITSEC defines six successful evaluation levels, from E1 to E6, representing ascending levels of security (E0 indicates inadequate security).

The *Canadian Trusted Computer Product Evaluation Criteria* (CTCPEC) publication is based on an approach combining the ITSEC and TCSEC approaches, but its requirements are closely compatible with individual TCSEC requirements [5]. Another document that combines the U.S. and European evaluation concepts is the draft *Federal Criteria for Information Technology Security* (FC), whose first public version was released in 1992. The FC never moved beyond the draft stage because the work was resumed under the umbrella of the Common Criteria project.

The security evaluation criteria developments mentioned in the previous paragraphs became input to the *Common Criteria for Information Technology Security Evaluation* [6]. This joint activity was initiated by the organizations that originally sponsored the TCSEC, ITSEC, CTCPEC, and FC. The aim was to develop a single set of security evaluation criteria for widespread use in the form of a proposal for the ISO as a contribution to the international standard under development. The following section explains the CC in more detail.

3.1 Common Criteria

The *Common Criteria for Information Technology Security Evaluation* (CC [6]) is the result of efforts to combine experience from a number of the existing European, U.S., and Canadian criteria (e.g., ITSEC, TCSEC, CTCPEC, and FC). Many different national organizations sponsor the CC project, such as CSE (Canada),[1] SCSSI (France),[2] BSI (Germany),[3] NLNCSA (The

---

1. http://www.cse-cst.gc.ca
2. http:// www.scssi.gouv.fr
3. http://www.bsi.bund.de

Netherlands),[4] CESG (United Kingdom),[5] NIST (United States),[6] NSA (United States),[7] and DSD (Australia).[8] An initial version (1.0) was released in January 1996, and the current version, 2.1 (as of December 2000), is also published as a series of ISO/IEC standards [3, 7, 8].

The CC is built around *security functional requirements* and *security assurance requirements*. The functional requirements [7] define the desired security behavior, whereas the assurance requirements [8] ensure that the alleged security measures are effective and implemented correctly [9]. A functional requirement is, for example, that all cryptographic keys on a smart card must be generated in accordance with a specified key generation algorithm (e.g., RSA) and that the keys must have a specified minimum length (e.g., 1,024 bits). An example of an assurance requirement is that each security mechanism implemented on the smart card (e.g., the key generation from the previous example) must be analyzed to show that it meets a specific minimum strength level and thus can resist certain security attacks. Regarding the attack potential from which a function must be protected, its strength of functions (SOF) can be defined as SOF-low, SOF-medium, or SOF-high. Obviously, highly critical parts of a system must satisfy SOF-high, which means that protection is provided against deliberately planned or organized breaches of system security by attackers having high attack potential [3].

A system under evaluation is referred to as the *target of evaluation* (TOE). A TOE is supposed to satisfy a set of security requirements specified in the form of a security target (ST). A more general way to specify security requirements is via a protection profile (PP), which represents an imple- mentation-independent set of security requirements that apply to a family of TOEs. For example, protection profiles are available for operating systems, firewalls, or smart cards.[9] Generally, three different types of evaluation are possible [8]:

1. A PP evaluation carried out against the evaluation criteria for PPs;

2. An ST evaluation carried out against the evaluation criteria for STs;

3. A TOE evaluation carried out against the evaluation criteria from [8] using an evaluated ST as the basis.

---

4. http://www.bsi.tno.nl

5. http://www.cesg.gov.uk

6. http://csrc.nist.gov

7. http://www.nsa.gov

8. http:// www.dsd.gov.au/infosec

9. http://csrc.nist.gov/cc/pp/pplist.htm

A protection profile contains an analysis of the TOE security environment. That is, it states

- Which assets should be protected;
- Which threats the TOE (i.e., the assets under protection) may be exposed to;
- Which assumptions are made about its secure usage;
- Which organizational security policies must be fulfilled.

As a result of the environment analysis, the protection profile defines the security objectives in such a way that they counter the threats and address the security policy and the assumptions. Finally, for each security objective, a set of security functional requirements and security assurance requirements is defined. This process is illustrated by Figure 3.1 [3]. Some examples of smart card protection profiles are given in the remaining sections in this chapter. PPs and STs can be developed by using the Common Criteria Toolbox,[10] a set of automated and freely available Java-based tools developed by the NIST.

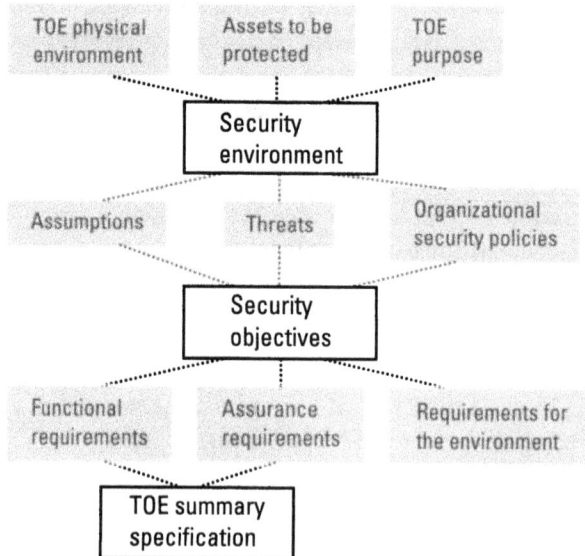

```
   TOE physical        Assets to be           TOE
   environment          protected           purpose


                      Security
                     environment


   Assumptions          Threats          Organizational
                                          security policies


                      Security
                     objectives


   Functional          Assurance         Requirements for
   requirements       requirements        the environment


                   TOE summary
                   specification
```

Figure 3.1 Common criteria: requirements and specifications.

---

10. http://niap.nist.gov/tools/cctool.html

The CC defines seven evaluation assurance levels (EALs), with EAL1 being the lowest level and EAL7 the highest. *Assurance* is a measure of confidence that a system meets its security objectives. The EAL represents the level of confidence one has regarding a security product. The higher the evaluation level, the more rigorous the evidence must be of the correct development process having been used and the higher the level of security protection required [8]:

- *EAL1 ("functionally tested"):* No interaction with the developer is required; the TOE functions in a way consistent with its documentation, and it provides useful protection against identified threats.

- *EAL2 ("structurally tested"):* This level is often applied to legacy systems or in cases for which information from the developer is insufficient; the TOE provides a low to moderate level of independently assured security.

- *EAL3 ("methodically tested and checked"):* The TOE is thoroughly investigated and provides a moderate level of independently assured security; no substantial reengineering is required.

- *EAL4 ("methodically designed, tested, and reviewed"):* This is the highest level that is likely to be economically feasible for an existing production line (e.g., smart cards for widespread use); the TOE provides a moderate to high level of independently assured security.

- *EAL5 ("semiformally designed and tested"):* TOE development is based on rigorous commercial development practices supported by moderate application of specialist security engineering techniques; the TOE provides a high level of independently assured security (*semiformal* means expressed in a restricted syntax language).

- *EAL6 ("semiformally verified, designed, and tested"):* The TOE protects high-value assets against significant risks; application of specialist security engineering techniques in a rigorous development environment is required.

- *EAL7 ("formally verified, designed, and tested"):* The TOE is developed for application in extremely high-risk situations or where the higher costs are justified; extensive formal analysis of the security functionality is required.

Each level is defined by a set of assurance *classes* and *families*. The following assurance classes are defined:

- *Configuration management (ACM):* Ensures that the TOE and documentation used for evaluation are the ones prepared for delivery, which means that TOE refinement and modification must take place in a controlled and disciplined way.

- *Delivery and operation (ADO):* Ensures that TOE security is not compromised during transfer, installation, start-up, and operation.

- *Development (ADV):* Requires that documentation be available for each development step, from the TOE summary specification in the security target down to the implementation, so that it can be determined whether the security requirements have been met.

- *Guidance documents (AGD):* Ensures that the user and administrator documentation is understandable and complete.

- *Life-cycle support (ALC):* Ensures that the life-cycle model is well defined for all development steps.

- *Tests (ATE):* Ensures through suitable testing procedures that the TOE security functional requirements are satisfied.

- *Vulnerability assessment (AVA):* Defines requirements that address exploitable vulnerabilities introduced in the construction, operation, misuse, or incorrect TOE configuration.

Each assurance class has several assurance families. For example, the vulnerability assessment class (AVA) includes families covering covert channel analysis (AVA_CCA), misuse (AVA_MSU), strength of TOE security functions (AVA_SOF), and vulnerability analysis (AVA_VLA). Each family further contains one or more assurance *components*. Finally, *assurance elements* deliver additional information for the assurance family (see Figure 3.2). They represent the actual evaluation steps. For example, the AVA_MSU family (misuse) contains components at three levels:

- *AVA_MSU.1, Examination of guidance:* There is no misleading, conflicting, or unreasonable guidance in the guidance documentation.

- *AVA_MSU.2, Validation of analysis:* In addition to AVA_MSU.1, the guidance documentation is analyzed by the developer.

- *AVA_MSU.3, Analysis and testing for insecure states:* In addition to AVA_MSU.2, the analysis is validated and confirmed by the evaluator.

Evaluation assurance level

↓

Assurance class

↓

Assurance family

↓

Assurance component

↓

Assurance element

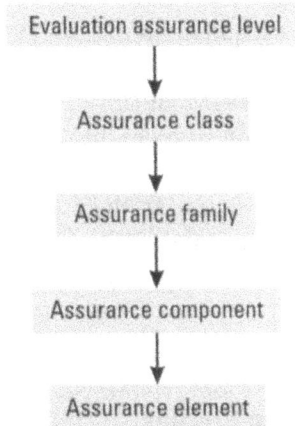

Figure 3.2 Common criteria: from evaluation level to action element (simplified).

An assurance element describes a detailed evaluation step, such as AVA_MSU.1.1D, which represents a developer action element: "The developer shall provide guidance documentation."

A mapping between the ITSEC and the CC evaluation levels is given in [10]. All previously mentioned national organizations sponsoring the CC project maintain a list of security products evaluated against the CC.

The following sections briefly introduce the most important of the currently available smart card profiles. The differences in their approaches to smart card security requirements specification can be seen by comparing threats and organizational security policies addressed by them. The main reason for the differences is that some profiles have been developed by integrated circuit (i.e., chip) or card manufacturers, and some by users. The integrated circuit contains in general a single chip with CPU, RAM, ROM, programmable nonvolatile memory (e.g., EEPROM), and (optionally) a cryptoprocessor.

3.2 Visa Smart Card Protection Profile

The TOE in the Visa Smart Card Protection Profile (VSCPP) is the integrated circuit, the card operating system, and applications of an integrated circuit card (i.e., smart card) [11]. The assets are not defined. The assurance level of the VSCPP is EAL4 augmented. This means that in addition to evaluation components required for EAL4, some additional or higher level evaluation families are required (in this case AVA_VLA.3, Vulnerability assessment—Vulnerability analysis—Moderately resistant, and ADV_INT.1, Development—TSF internals—Modularity). The VSCPP focuses on the card usage and card termination stages of the card life cycle from ISO 10202 (see Section 1.4), which also means that this card life cycle is a requirement of the VSCPP. The threats at these stages must be addressed at the manufacturing and card preparation stage. Because it was developed by a card manufacturer, this protection profile relates mostly to the manufacturing process.

3.3   Eurosmart Protection Profiles

The Eurosmart (European Smart Card Industry Initiative) Security Working Group has published three smart card–related protection profiles. The participants in this working group are representatives of many chip and card manufacturers (Giesecke & Devrient, Schlumberger, Gemplus, Philips, Infineon, and many others).

The most general protection profile addresses the smart card IC with embedded software [12]. The embedded software may be in any part of the nonvolatile memories and includes the basic software (i.e., operating system, interpreters) and the application software. The asset under protection is the TOE itself:

- The IC specifications, design, development tools, and technology;

- The IC dedicated software (for testing purposes);

- The smart card–embedded software including specifications, implementation, and related documentation;

- The TOE application data (i.e., IC system data, initialization data, personalization data).

The assurance level is EAL4 augmented, with the following augmentations:

- ADV_IMP.2, Development—Implementation representation— Implementation of the TOE security functions;

- ALC_DVS.2, Life-cycle support—Development security—Sufficiency of the security measures;

- AVA_VLA.4, Vulnerability assessment—Vulnerability analysis—Highly resistant.

The second protection profile is for the intersector electronic purse and purchase device [13], which consists of both parts of the TOE. This protection profile identifies the threats, security policies, and objectives specific to this type of payment mechanism. The intersector electronic purse (IEP) consists of an IC with embedded software that is compliant with the previously described general protection profile. The assurance level is the same as in that profile. The IEP is prepaid, reloadable, optionally anonymous, and interacts with the purchase device, also a part of the TOE. The purchase device (PD) is installed at the service provider to accept payments from an IEP. The PD may be a part of the terminal or a server.

The third protection profile addresses smart cards with a secure multiple-application platform [14]. *Secure* in this context also means that the applications cannot access each other's code or data unless explicitly allowed. In addition to the usual smart card IC parts, this smart card's IC also contains an application system interface (e.g., loader, virtual machine, card administrator) and native or interpreted software at the application level. Consequently, the assets additionally include application software loaded onto the IC, data used and manipulated by the application, and card resources (memory space and computation power). The assurance level is very high (i.e., EAL5 augmented). The augmentations are ALC_DVS.2 and AVA_VLA.4 (see the general Eurosmart protection profile given earlier).

3.4 Smart Card Security User Group's Protection Profile

The Smart Card Security User Group's Protection Profile (SCUGPP) is a user-oriented protection profile [15]. It is application independent, which means that for a complete evaluation it should be supplemented with an application-specific protection profile. The primary asset is the user data to be protected, that is:

- Cardholder data in support of TOE functions;
- Application code added to the TOE to implement additional functionality;
- Security attributes, authentication data, and access control list;
- Cryptographic keys for TOE security processes.

The assurance level is EAL4 augmented with the same augmentations as in the VSCPP (see Section 3.2). The TOE comprises the integrated circuit, the operating system, and the mechanisms that allow communication with the outside world. The TOE can establish a secure channel to a trusted source for application loading or execution of privileged commands.

3.5 Secure Signature-Creation Device Protection Profile

The Secure Signature-Creation Device Protection Profile (SSCD-PP), a draft version solicited for public comments, was prepared for the European Electronic Signature Standardisation Initiative (EESSI) by CEN/ISSS area F [16]. The SSCD-PP defines security requirements for signature applications for any type of signature-creation device, not necessarily smart cards, in accordance with the EU directive on electronic signatures [17].

The assets to be protected cover the signature-creation data (SCD, i.e., the private key), the signature-verification data (SVD, i.e., the public key), the document to be signed, and the authentication data (e.g., the PIN). The TOE (the secure signature-creation device) contains the SCD, which must be unique for the signatory who owns the TOE. It must be impossible to obtain the SCD from the TOE. One of the policies requires the use of the qualified certificate defined in [17] containing the name of the signatory and the SVD matching the SCD implemented in the TOE. The signature-creation function on the TOE can be activated only if the signatory presents valid authentication data (e.g., the PIN).

Two documents have been forwarded to the EESSI Steering Committee, one requiring the assurance level to be EAL4, and one requiring it to be EAL4 augmented, the augmentations being AVA_MSU.3 (Vulnerability assessment—Misuse—Analysis and testing of insecure states) and AVA_VLA.4 (Vulnerability assessment—Vulnerability analysis—Highly resistant). The reason for having two different assurance levels was that a number of industry representatives argued against the originally proposed augmentations. The decision—EAL4 or EAL4 augmented—should be made by the Electronic-Signature Committee (see Article 9 in [17]).

References

[1] National Computer Security Center, *Department of Defense Trusted Computer System Evaluation Criteria (TCSEC or Orange Book)*, DoD 5200.28-STD, Washington, D.C.: NCSC, 1983; available at http://csrc.ncsl.nist.gov/secpubs/rainbow/std001.txt.

[2] "Computer Security Evaluation FAQ, Version 2.1," Feb. 1998; available at http://www.landfield.com/faqs/computer-security/evaluations.

[3] International Organization for Standardization, *Information Technology—Security Techniques—Evaluation Criteria for IT Security—Part 1: Introduction and General Model*, ISO-IEC 15408-1, Geneva: ISO, 1999.

[4] Commission of the European Communities, *Information Technology Security Evaluation Criteria (ITSEC)*, Office for Official Publications of the European Communities, Version 1.2, June 1991; available at http://www.cordis.lu/infosec/docs/itsec.zip.

[5] Canadian System Security Center, *The Canadian Trusted Computer Product Evaluation Criteria*, Version 3.0, Jan. 1993; available at ftp://ftp.cse-cst.gc.ca/pub/criteria/ CTCPEC/CTCPEC.ascii.

[6] International Common Criteria home page, 2000; available at http://www.common criteria.org/.

[7] International Organization for Standardization, *Information Technology—Security Techniques—Evaluation Criteria for IT Security—Part 2: Security Functional Requirements*, ISO-IEC 15408-2, Geneva: ISO, 1999.

[8] International Organization for Standardization, *Information Technology—Security Techniques—Evaluation Criteria for IT Security—Part 3: Security Assurance Requirements*, ISO-IEC 15408-3, Geneva: ISO, 1999.

[9] Common Criteria Project Group, "Common Criteria Version 2: An Introduction," Oct. 1998; available at http://csrc.nist.gov/cc/info/cc_brochure.pdf.

[10] Bundesamt fuer Sicherheit in der Informationstechnik, "Application Notes and Interpretation of the Scheme (AIS): ITSEC to CC Mappings with Specific Attack Potential," AIS 27, Version 1, April 2000; available at http://www.bsi.bund.de/.

[11] Visa International Service Association, "Visa Smart Card Protection Profile," Draft Version 1.6, May 1999; available at http://www.visa.com/nt/chip/drpp-v.pdf.

[12] European Smart Card Industry Initiative, "Protection Profile Smart Card Integrated Circuit with Embedded Software," PP/9911, Version 2.0, June 1999; available at http://www.eurosmart.com/Activities/DownloadArea/Files/CPP9911.pdf.

[13] European Smart Card Industry Initiative, "Protection Profile Intersector Electronic Purse and Purchase Device," PP/9909, Version 1.2, Feb. 1999; available at http://www.eurosmart.com/Activities/DownloadArea/Files/CPP9909.pdf.

[14]   European Smart Card Industry Initiative, "Protection Profile Smart Card IC with Multi-Application Secure Platform," PPnc/0001, Version 1.0, Jan. 2000; available at http://www.eurosmart.com/Activities/DownloadArea/Files/pp0001.pdf.

[15]   Smart Card Security User Group, "Smart Card Protection Profile," Draft Version 2.0, May 2000; available at http://csrc/nist.gov/cc/sc/sclist.htm-SCSUG-PP.

[16]   European Committee for Standardization, Information Society Standardization System, "Protection Profile—Secure Signature-Creation Device," Version 1.0, Nov. 2000.

[17]   "Directive 1999/93/EC of the European Parliament and of the Council of 13 December 1999 on a Community Framework for Electronic Signatures," *Official Journal* L 013, Jan. 19, 2000, pp. 12–20; available at http://europa.eu.int/eur-lex/en/lif/dat/1999/en_399L0093.html.

4

File Structure and Commands

The most important data structures and commands for integrated circuit cards with contacts are standardized in ISO/IEC 7816-4 and ISO/IEC 7816-8 (see Table 1.1). While Part 4 describes the basic data structures and commands, Part 8 is dedicated to enhanced cryptographic functions. Other standards are application specific, for example, the EN 726-3 for GSM SIM[1] cards, the EN 1546 for electronic purses, or the EMV specifications for debit and credit cards.

4.1 File Structure of Integrated Circuit Cards

The ISO/IEC 7816-4 file organization follows a tree structure as in desktop computer operating systems. There are elementary files (EF) that contain data and dedicated files (DF) that group other files. However, the DFs do not generally contain a directory of all subordinate files. The root DF is called the master file (MF). Figure 4.1 gives an example of a valid smart card file structure.

Files are selected by their 2-byte file identifier (FID). The MF always has the FID = 3F 00 H and is automatically selected whenever the card is reset. FIDs must be unique within each DF. An EF may also have a short file identifier, which is only 5 bits long and may have a value from 1 to 30. This

---

1. Global System for Mobile (GSM) communications SIM.

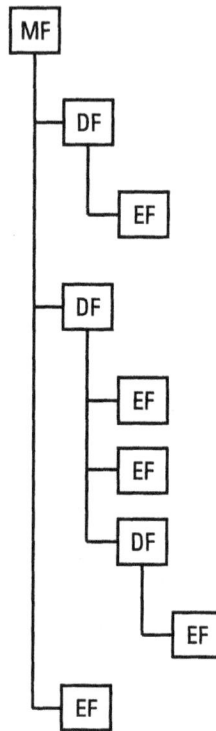

Figure 4.1 Example of a smart card file structure.

is used for implicit file selection, which makes it possible to use a number of card commands directly without explicitly selecting a file beforehand.

A DF can have a 1- to 16-byte name in addition to its FID. This name should be unique within the card and may be a text string or a registered application identifier (AID) as defined in ISO/IEC 7816-5 (see Table 1.1). An AID consists of a 5-byte registered identifier (RID), denoting the (juristic) person who has applied to register an application, and a proprietary application identifier (PIX), which is up to 11 bytes long and is chosen by the person who registers the application.

EFs can have different internal structures:

- Transparent EFs contain unstructured data.

- Linear fixed EFs are divided into records of fixed length. A number ranging from 1 to FE H is assigned to each record.

- Linear variable EFs are divided into records of variable length. Again, a number ranging from 1 to FE H is assigned to each record.
- Cyclic fixed EFs are like linear fixed EFs divided into records of fixed length. They have a certain record numbering scheme that makes it easy to append new records infinitely because the oldest record is automatically overwritten.

4.2 Command Structure of ICCs

Integrated circuit cards are passive elements that receive commands from the terminal and respond to them. Every smart card command or response is coded in an application protocol data unit (APDU). The structure of a command APDU with its mandatory header and its conditional body is shown in Figure 4.2. The class byte (CLA) indicates the general command type. For example, ISO commands have CLA = 0x H, GSM commands have CLA = Ax H, and proprietary commands have CLA = 8x H. The letter "x" defines the applied secure messaging format (see Section 4.4).

The instruction byte (INS) specifies a certain command within a class. The parameter bytes (P1, P2) are used to choose between different options a command may offer. The body is a concatenation of the length of the command (L_c), the number of data bytes, and the length of the expected response data (L_e). It is important to note that L_c is not calculated over the whole command but only over the data. Depending on the operating system, both lengths may be coded in 1 or 3 bytes. The latter option, known as extended length coding, uses the first byte as an escape character and two more bytes for the length value. Thus a maximum length of 65,536 data bytes can be reached.

The structure of a response APDU (see Figure 4.3) is somewhat simpler. It consists of a conditional body and a status word (SW1, SW2) that is 2 bytes long and also called a return code. If the command is processed without any restrictions, the card will usually respond with the status word 90 00 H.

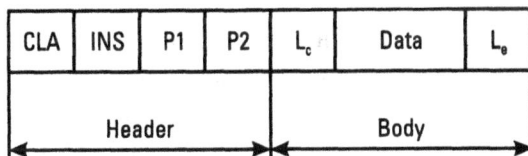

CLA	INS	P1	P2	L_c	Data	L_e
Header				Body		

Figure 4.2 Command APDU structure.

Data	SW1	SW1
Body	Trailer	

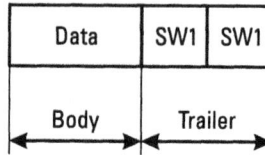

Figure 4.3 Response APDU structure.

In any other case, depending on the instruction, different errors or warnings are indicated through certain values of the status word.

4.3 Examples of Smart Card Commands

The most important smart card commands that will be needed later in the book are discussed in more detail here. Normally, a series of card commands starts with a number of SELECT FILE instructions to access the correct EF. Figure 4.4 shows some specific options of this command.

As mentioned earlier, files may be explicitly selected by their FID. For DFs, selection by name is optional, in which case the DF name need not be fully stated in the data of the SELECT FILE command, but may be truncated. This is called *partial selection* and uses the options of selecting the first, last, next, or previous file where its name starts with the stated bytes. Files can also be selected by stating a path. The path is simply a concatenation of FIDs starting from the MF or the current DF. The FID of the starting point is not included in the path.

00 H	A4 H	P1	P2	L_c	Data	L_e

P1 options:	P1 = 00 H	Select by FID
	P1 = 04 H	Select DF by name
	P1 = 08 H	Select from MF by path
	P1 = 09 H	Select from current DF by path
P2 options:	P2 = 00 H	First occurrence
	P2 = 01 H	Last occurrence
	P2 = 02 H	Next occurrence
	P2 = 03 H	Previous occurrence

Figure 4.4 SELECT FILE command.

Figure 4.5 shows the response to the SELECT FILE command. Depending on the operating system, different formats of the file control information (FCI) may be given in the response. The FCI contains information such as FID, file type, file size, record length, and similar. Some common return codes indicate parameter errors or that the card was unable to find the file.

The VERIFY command is used to compare verification data like passwords or PINs to the corresponding reference values stored in the card. Different rights to read or write a certain file may be bound to such a password verification or a cryptographic authentication (see Section 4.4). Figure 4.6 shows as the only option a choice of global reference data, for example, a card's master PIN, or application-specific reference data, for example, a DF's PIN. The least significant 5 bits of P2 denote the reference data number inside the card.

The length L_e is left empty, as the response to the VERIFY command contains no data (see Figure 4.7). If the verification fails because of incorrect verification data, the card will in most cases indicate the number of remaining tries in the status word. After the last try the card will block the verification of these data.

READ RECORD is another command that is found in many applications. An overview of its options is given in Figure 4.8. The parameter bytes

FCI data	SW1	SW2

SW coding: SW = 6A 82 H   File not found

SW = 6A 86 H   Incorrect P1, P2

SW = 6A 87 H L_c inconsistent with P1, P2

Figure 4.5 SELECT FILE response.

00 H	20 H	00 H	P2	L_c	Verif. data

P2 options: P2 = 00 H   No further information

P2 = 0X H   Global reference data

P2 = 8X H   Specific reference data

Figure 4.6 VERIFY command.

SW1	SW2

SW coding: SW = 63 00 H   Verification failed, no further information

SW = 63 CX H   Verification failed, X retries left

SW = 69 83 H   Authentication method blocked

SW = 6A 86 H   Incorrect P1, P2

SW = 6A 88 H   Reference data not found

Figure 4.7 VERIFY response.

0X H	B2 H	P1	P2	L_e

P1 options: P1 = 00 H   Current record

P1 = XX H   Record number

P2 options: P2 = 0X H   Current EF + record options

P2 = XX H   Short FID + record options

Figure 4.8 READ RECORD command.

allow direct selection of records by their number or browsing through the records by setting P1 to "current" and using the options of P2 to select the next, previous, first, or last record. Another option is to read just one record or all records. The upper 5 bits of P2 can contain a short FID to perform an implicit file selection.

Because READ RECORD does not need to send any data, L_c and the data field are left empty. L_e is needed, of course, but may be zero to indicate that all data shall be read. Figure 4.9 shows the corresponding response. In the case of an incorrect L_e, the card may indicate the correct value.

Other standardized smart card commands are listed in Table 4.1.

4.4 Cryptographic Authentication and Secure Messaging

The principle of cryptographic authentication was explained in Chapter 2. The relevant commands according to ISO/IEC 7816-4 are INTERNAL AUTHENTICATE and EXTERNAL AUTHENTICATE:

Data	SW1	SW2

SW coding: SW = 62 82 H   End of record before L_e bytes

SW = 67 00 H   Wrong length

SW = 6C XX H   Wrong length, expected L_e = XX

SW = 69 81 H   Wrong options for this file structure

SW = 69 82 H   Security status not satisfied

SW = 6A 81 H   Function not supported

SW = 6A 82 H   File not found

SW = 6A 83 H   Record not found

Figure 4.9 READ RECORD response.

Table 4.1
Smart Card Commands

INS	Command Name	Purpose
82	EXTERNAL AUTHENTICATE	Authentication
84	GET CHALLENGE	Return a random number
88	INTERNAL AUTHENTICATE	Authentication
B0	READ BINARY	Read data from transparent EFs
C0	GET RESPONSE	Receive response APDUs
CA	GET DATA	Read a specific data object
D0	WRITE BINARY	Write data to transparent EFs
D2	WRITE RECORD	Write data to linear or cyclic EFs
D6	UPDATE BINARY	Update data in transparent EFs
DA	PUT DATA	Write a specific data object
DC	UPDATE RECORD	Update data in linear or cyclic EFs
E2	APPEND RECORD	Append a record to linear or cyclic EFs

- INTERNAL AUTHENTICATE is used for an authentication of the integrated circuit card to the terminal or to the background system. With this command, the card receives a challenge and the identifier of the key it shall use to encrypt the challenge. In the response it delivers the result of the encryption, which is then checked by the terminal.

- EXTERNAL AUTHENTICATE is used for an authentication of the terminal to the integrated circuit card. The terminal encrypts a challenge and sends the result together with the identifier of the key to the card within this command. The result is then checked by the card.

In the case of EXTERNAL AUTHENTICATE, the challenge shall be provided by the card. This is performed with the command GET CHALLENGE.

Mutual authentication can be achieved by use of EXTERNAL AUTHENTICATE followed by INTERNAL AUTHENTICATE. To speed up this procedure and to improve security further by not transmitting clear data between card and terminal, ISO/IEC 7816-8 defines the command MUTUAL AUTHENTICATE. The most important command of the ISO/IEC 7816-8 standard is PERFORM SECURITY OPERATION, which is able to calculate hash values, MACs, and digital signatures and to encrypt and decrypt data with both symmetric and asymmetric algorithms, if supported by the card operating system.

Secure messaging is a name for all mechanisms that provide authenticity, confidentiality, or both of the communication between the card and the terminal or the background system. The type of secure messaging is coded in the upper 2 bits of the lower nibble of the class byte. For instance:

- CLA = x0 H indicates no secure messaging.
- CLA = x4 H indicates proprietary secure messaging.
- CLA = x8 H indicates ISO-compliant secure messaging without authenticated command header.
- CLA = xC H indicates ISO-compliant secure messaging with authenticated command header.

Various cryptographic mechanisms can be used for secure messaging, ranging from a simple MAC to digital signature plus encryption using asymmetric cryptography and certificates. Two terms are widely used in this context: *authentic mode* refers to adding a MAC only, and *combined mode* refers to additionally encrypting the resulting data block. Before cryptographic mechanisms are applied, some padding may be needed to make the length of the data block valid. Note that when ISO-7816-compliant secure messaging is used, all data elements must be coded in BER-TLV.[2]

An authentic mode command with authenticated command header looks then like the example given in Figure 4.10. The header and the data block(s) of the command are separately padded. Of course, the length of the command is increased through this process. The resulting command consists of the original header and a new data block containing the BER-TLV–coded original data and the added MAC, which is BER-TLV coded.

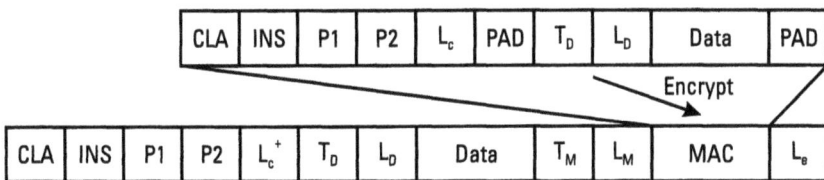

Figure 4.10 Command with secure messaging in authentic mode.

---

2. Basic encoding rules (BER) according to ISO/IEC-8825 for tag-length-value (TLV) data structures compliant with Abstract Syntax Notation One (ASN.1) according to ISO/IEC-8824.

5

ISO 7816 Smart Card Communication

ICCs with contacts are standardized in ISO/IEC 7816 (see Table 1.1), where Part 1 specifies the physical characteristics and Part 2 the dimensions and location of the contacts. Figure 5.1 shows the allocation of the eight contacts.

VCC and GND are needed for the power supply, which is typically 5V or 3V DC. VPP was needed in earlier years for supply of a separate EEPROM programming voltage. As of 2001, this voltage is generated internally and the contact is no longer used. The external clock is connected to CLK, and RST allows resetting of the integrated circuit. All data are transmitted via a single input-output (I/O) contact, hence only half-duplex protocols are possible. These are simply master/slave protocols where the terminal is the master and the card is the slave. This means that only the terminal is permitted to initiate the communication and the card has merely to respond. Because there are two more RFU contacts on the IC that are not used

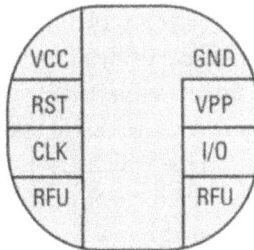

Figure 5.1 ICC contacts.

anymore, future use of those contacts and the implementation of full-duplex protocols are, in principle, possible.

At the physical layer of communication, ISO/IEC 7816-3 specifies an asynchronous, serial protocol (see Table 1.1). Every byte is coded with 1 start bit, the 8 data bits, 1 parity bit, and 1 or 2 stop bits, which are named *guard time*. Every bit is transformed to a voltage level by either direct or inverse convention. With direct convention, every logical 1 is coded as a high-voltage and every logical 0 is coded as a low-voltage level. With inverse convention, every logical 0 is coded as a high and every logical 1 is coded as a low-voltage level.

The transmission speed (baud rate) depends on the external clock provided by the terminal. This clock frequency is divided by a certain factor, typically 372 or 512, to obtain the baud rate. For the basic baud rate of 9,600 bps, terminals typically apply the clock frequencies of 3.5712 or 4.9152 MHz. Yet baud rates up to 312,500 bps are possible. Because the clock depends on an external source, there is a need for another measure to be able to specify the timing of a smart card. The elementary time unit (etu) is defined as the duration of the transmission of 1 bit.

5.1 Answer to Reset

After a cold or warm reset activated by the RST signal, every ISO 7816-compliant integrated circuit card must send the *Answer to Reset* (ATR) via the I/O contact. During this ATR transmission, 1 etu must be equal to $372/f$, where f is the applied clock frequency. The ATR consists of a number of bytes that indicate the capabilities of the integrated circuit. Figure 5.2 shows its general structure.

The initial character TS has only two allowed values, which represent direct or inverse convention, respectively. The format byte T0 indicates the structure of the ATR. Its upper 4 bits show the presence of subsequent optional interface characters TA(1) to TD(1), and its lower section shows the number of historical characters, which are optional as well. The interface character TA(1) contains the coded clock rate conversion factor F (ranging from 372 to 2048) and the coded baud rate adjustment factor D (ranging from 1 to 32). After the ATR, those factors can be applied and 1 etu will then be equal to $(F/D)/f$. The second interface character TB(1), if present, shows the voltage and the maximum current for VPP. TC(1) contains the extra guard time N. This tells the terminal to wait additional N etus after sending one character before sending the next one.

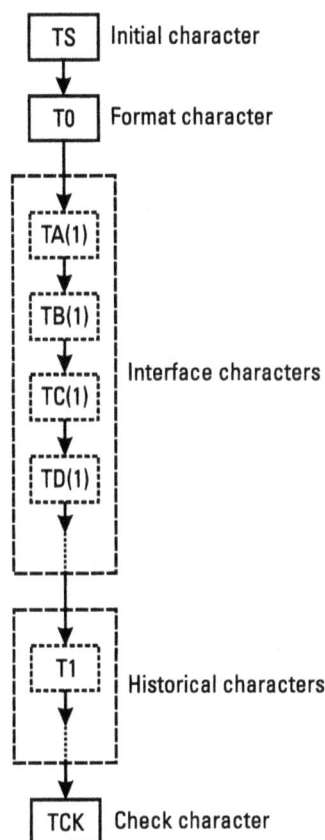

Figure 5.2 ATR structure.

The fourth interface character TD(1) indicates the presence of further interface characters (as in T0) and the transmission protocol parameter T in its lower nibble. The following values of T are defined:

- T = 0 is a half-duplex and byte-oriented protocol. It was the first smart card protocol and is still widely used.
- T = 1 is a half-duplex and block-oriented protocol. It is more advanced than T = 0 and offers a clearly layered architecture as well as good error recognition and recovery techniques.
- T = 2 and T = 3 are full-duplex protocols that have not yet been standardized.
- T = 4 to T = 13 are reserved for future use (RFU).

- T = 14 indicates proprietary protocols not standardized by ISO/IEC. Different T = 14 protocols are in use today, for example, for telephone cards and pay TV cards.

- T = 15 does not refer to a particular protocol. This value is used in interface characters TD(i) to declare global interface characters.

The interface character TA(2) selects the specific mode where a specified protocol with specified parameters is directly applied after the ATR, or the negotiable mode where a further protocol and parameters selection (PPS) may be performed by the terminal. The interface character TB(2) can show a VPP voltage that is not normally used today. TC(2) can contain a parameter specific for T = 0. TD(2) indicates further interface characters TA(i) to TD(i), which are valid for protocol T in TD(2).

Up to 15 optional historical characters contain information provided by the operating system developer. Examples are the chip type, the name of the operating system or application, a version number, the life-cycle state (operating system completed or not completed), or the name of the card manufacturer. This information is usually coded in ASCII.

The last element of the ATR is the check character TCK. It is calculated in such a way that an exclusive-or operation on all ATR bytes including TCK gives the result zero, with which the terminal can verify the integrity of the ATR.

Figure 5.3 gives an example of an ATR and its interpretation. This card with the operating system Starcos SPK 2.3 uses direct convention, no VPP voltage, and the protocol T = 1.

| 3B | B7 | 94 | 00 | 81 | 31 | FE | 65 | 53 | 50 | 4B | 32 | 33 | 90 | 00 | D1 |

Historical characters: „SPK23" 90 00

TB(3): specific T = 1 parameter   TCK
TA(3): specific T = 1 parameter
TD(2): TA(3) and TB(3) follow, T = 1 is supported
TD(1): TD(2) follows, T = 1 is supported
TB(1): No VPP voltage
TA(1): F = 512, D = 8
T0: TA(1), TB(1) and TD(1) follow, 7 historical characters
TS: Direct convention

Figure 5.3 ATR example.

5.2 T = 1 Protocol

The T = 1 protocol is the most important protocol today for high-security smart cards, particularly banking cards and digital signature cards. The execution of the protocol starts after the ATR and any PPS phase, where the baud rate can be changed. As explained earlier, during the half-duplex communication protocol, the terminal sends commands to which the card responds. Besides these *information blocks* containing APDUs, *receive ready blocks* are used for acknowledgments, and *supervisory blocks* transport control information, such as a waiting time extension request and response. By use of a waiting time extension request, a card can indicate that it needs more time for the preparation of the response to a command. After the waiting time extension response from the terminal, the card can send the response to the original command.

The general block format is shown in Figure 5.4. The node address byte (NAD) contains the source and destination address of the block. In many cases there is only one logical connection between a card and a terminal; hence, there is no need to distinguish source and destination addresses, and the NAD byte may be set to 0. The protocol control byte (PCB) indicates the type of block: information block, receive ready block, or supervisory block. A PCB of an information block contains a 1-bit sequence number. The length byte (LEN) contains the length of the information field.

There may be different maximum information field sizes for the card and the terminal. If longer data have to be transmitted, a chaining mechanism can be applied: A bit in the PCB of an information block indicates that more data are to follow; the communication partner then sends a receive ready block containing the sequence number of the expected next information block. The error detection code (EDC) is calculated over the prologue and the information data fields. Error handling mechanisms include retransmission of blocks and special supervisory blocks for resynchronization.

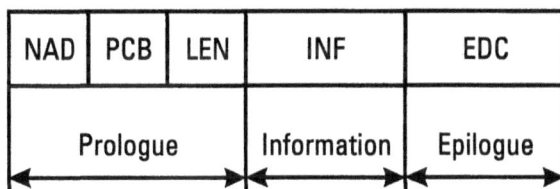

Figure 5.4 T = 1 block format.

6

Card Readers, Card Terminals, and Related Technologies

6.1 Card Readers and Card Terminals

Different names have been given to devices that have the ability to communicate directly with an integrated circuit card. General terms are card accepting device (CAD) or interface device (IFD). These devices can be divided into two classes:

1. Card readers do nothing more than supply the card with power and clock and execute the communication protocol. Usually they are connected to a personal computer or other background system. Depending on the power consumption of the card reader itself, it may need its own power supply, but some types are supplied with power via the serial interface of a personal computer.

2. Card terminals have additional capabilities to allow user interaction, such as a liquid crystal display (LCD) and a numerical keypad. One special type of terminal is the value checker, which is a battery-powered device that allows the user to read certain data from a card, for example, the balance of an electronic purse or diverse transaction log files.

Throughout this book, the word terminal is used because e-payments normally require interaction with the user or the merchant.

Figure 6.1 shows the internal structure of a card terminal. The terminal software is usually held in an EPROM. I/O units are needed for the display, the keypad, and the communication with the background system and the card. If the terminal has to perform sensitive operations, these will preferably be done by a special security module (SM). Because the security module should be tamper-proof, it will in many cases be a smart card itself. Terminals without security modules should never contain secret keys, because an attacker would be able simply to steal a terminal and read out the keys from it instead of having to attack a smart card. Because terminals usually have to perform transactions with all types of issued cards, they often need a master key, which is even more dangerous. This makes terminals without security modules unsuitable for performing operations of symmetric cryptography; they should only be used for operations with public keys of asymmetric cryptography, that is, signature verification and asymmetric encryption (but not digital signature).

A terminal's online and offline capabilities are related attributes. A terminal will be capable of performing secure offline transactions if an appropriate security module is installed or security can be achieved with public key operations. If not, an online capability is needed where the background system, which is equipped with all relevant keys, can perform the transaction with the card. Of course, also required for performing transactions offline is for the terminals to have some way to exchange data with the background system. This can be achieved via a periodically established online connection

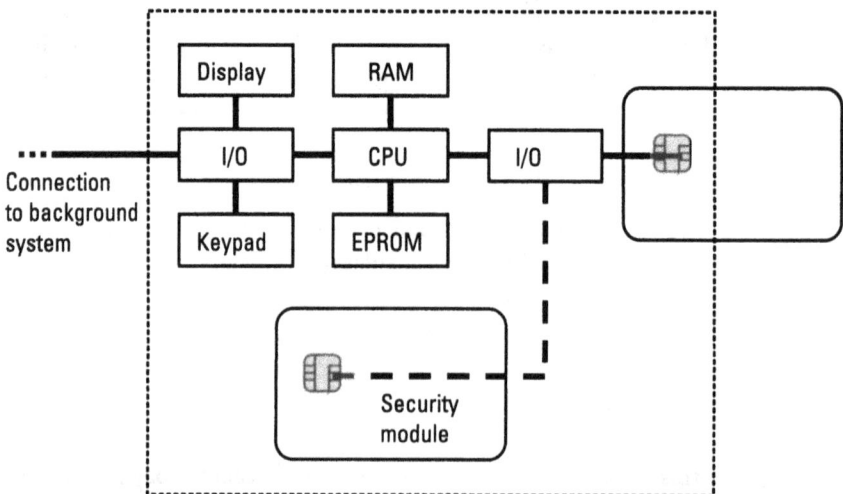

Figure 6.1 Internal structure of a card terminal.

or by data transport via a physical data medium. This medium may again be a smart card, but floppy disks and other media are also suitable if the data to be transported have been cryptographically secured by a trusted entity.

Technical parameters of terminals include the following:

- Voltage and maximum power supply to the card;
- Clock frequency;
- Power supply and power consumption of the terminal;
- For ICCs with contacts, contact material and landing or sliding contacts (the latter usually cause more abrasion of the cards' contacts);
- For contactless ICCs, communication range;
- Supported communication standards and communication protocols to the card for example, ISO 7816-3 T = 1 or memory card protocols like $S = 8$ (also called I^2C);
- Speed of data transmission to the card;
- Connection and communication protocol to the background system;
- Speed of data transmission to the background system;
- Number of slots for security modules;
- Dimensions and weight;
- Temperature range.

6.2 Related Technologies

Smart cards can generally be used in various types of devices (mobile phones, personal digital assistants, set-top boxes, and others), assuming the device is equipped with a suitable card reader. More information about deployment of the Java Card technology is given in Chapter 9.

This chapter briefly presents another technology that is more and more frequently mentioned in connection with smart cards, especially in the United States, namely, universal serial bus (USB). USB is a system that provides two-way communication between multiple peripheral devices and a host computer (e.g., a PC). A single USB interface is attached to the motherboard. A root hub with up to seven additional hubs can be integrated into the main interface or externally connected with a cable. Each of the seven hubs on the root hub can in turn be connected to seven more hubs, and so on, yielding a maximum of 127 ports. USB is already supported by a number of operating systems [1]. According to some estimates, more than 50% of all PCs in the world will be USB enabled by the end of 2001.

The importance of the USB technology for smart cards is that USB offers a simple way to "plug in" a card to a personal computer (i.e., no special card reader is necessary). In other words, the connection between the card and the PC is established via the USB port. However, the interface electronics normally resident in a traditional smart card reader must be added to the chip on the card [2].

A small-format card within a token that plugs directly into a USB port is usually referred to as a *USB token.* Industry has made proposals that would standardize the USB protocol for the next generation of SIM cards for third-generation mobile phone systems because USB offers higher data transfer rates (full speed at 12 Mbps, with low-speed devices using a 1.5 Mbps sub-channel). USB tokens can be used to encrypt data and store user passwords, and as private keys and digital certificates for authentication solutions for local area networks, virtual private networks, e-commerce, and mobile computing, to name a few examples [3].

References

[1]   Quatech, "USB Overview," 2000; available at http://www.quatech.com/Application_Objects/FAQs/comm-over-usb.htm.

[2]   Schlumberger, "Schlumberger Opens the Fast Track to Secure E-Business with Cryptoflex E-Gate: USB Smart Card Innovation," Press release, Nov. 16, 1999; available at http://www.1.slb.com/smartcards/news/pr99/second/sct_egate1611.html.

[3]   Aladdin Knowledge Systems, "Aladdin Knowledge Systems Introduces Next-Generation, Secure USB Authentication Token," Press release, May 2000; available at http://www.esafe.co.za/press_forum/secure.html.

7

Debit and Credit Cards

Debit and credit card transactions are in principle based on a rectangular relationship (see Figure 7.1). The customer receives the card from the issuer and uses it to spend money at the merchant's. The merchant performs an electronic transaction with the acquirer who in turn involves the issuer. In the case of a debit transaction, the amount spent is deducted from the customer's bank account without any delay beyond the actual processing time. This is also called "pay now," while the case of a credit transaction is called "pay later."

For decades electronic transactions with debit and credit cards were based solely on magnetic-strip technology. Local or national systems with

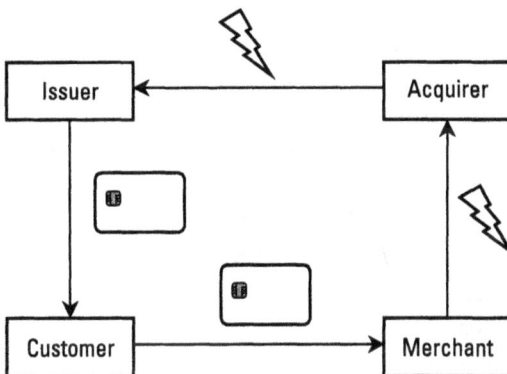

Figure 7.1 The principle behind debit and credit card transactions.

automatic teller machines (ATM) and point-of-sale (POS) payment terminals were set up and connected to each other to allow international transactions. Although those systems provided sufficient functionality, their security level allowed attacks such as copying of entire cards, including all magnetic-strip data.

By the early 1990s, the amount of fraud in many countries had reached such a high level that banks and credit card organizations were looking hard for serious measures to prevent it. Despite the higher costs, the technology was changed from magnetic-strip cards to smart cards because this solution provided the highest security level. Furthermore, smart card technology allowed the addition of new features. POS transactions were usually performed with an online connection to some background system. Because of the greater security provided by smart cards, offline transactions were now also possible. The POS terminal only had to go online periodically, typically once a day, to upload transaction data and download new "blacklists" of suspicious cards. This reduced the merchant's costs and helped to convince more merchants of the benefits of POS terminals.

Another feature that has often been implemented is the electronic purse. Purse payment transactions are not related to a customer's bank account. In this case, a certain amount of money is stored directly in the chip and can usually be spent without the need to enter a PIN. During the transaction, the amount is transferred to the merchant's card, which is plugged into the terminal. Most purses can be loaded from a bank account at ATMs and some systems allow direct purse-to-purse transactions as well.

Table 7.1 gives an overview of the most widespread smart card–based banking card schemes providing electronic purses. We can see that Austria had the first country-wide electronic purse, and that Germany has the highest number of cards issued as of 2001. Unfortunately, all of these different solutions were specified and implemented without concern for interoperability. Hence, international use of electronic purses is not possible, and international debit and credit transactions still rely on magnetic-strip technology. This is even more inconvenient to customers in countries of the European Union (EU) that have adopted the Euro as their common currency and will use the same coins and banknotes beginning in 2002. Many people will then hold Euros in their electronic purses, but unlike the coins and banknotes, the local "electronic Euros" cannot be used in other countries of the EU.

As we know, different solutions to the interoperability problem have been proposed. Sometimes individual companies have grown strong enough to "rule the world" such that their own products become the "standard." This is not only a source of anger to competitors but also leads to weaknesses

Table 7.1
Banking Card Schemes with Electronic Purses

Country	Electronic Purse	Country-Wide Roll-Out	Number of Cards as of 2001 (millions)
Austria	Quick	1995	4
Belgium	Proton	1997	7
Germany	Geldkarte	1997	50
The Netherlands	Chipknip	1997	12
Spain	Visa Cash	—	4
Switzerland	Cash	1997	3

arising from the one-sided view. Sometimes the need for standardization is understood early enough for a standardization committee to be founded in which everyone can be represented and can contribute. In such a situation, proposals are carefully discussed before an international standard is created,[1] which may take some time. In the case of debit and credit cards, a hybrid approach was chosen. The three most important organizations, Europay, MasterCard, and Visa (EMV), established a common working group that has come up with the joint *EMV Integrated Circuit Card Specifications for Payment Systems* defining smart card–based debit and credit transactions.[2]

7.1 Relevant Specifications

Work on the EMV specifications began in 1994, and several versions have been published over the years, with the *EMV '96 Integrated Circuit Card Specification, Version 3.1.1,* being the first stable one. By the end of 2000 the

---

1. Unfortunately, the existence of a standard does not always ensure the interoperability of products of different vendors, because not every implementation detail is covered in the standard. This explains the growing number of certification laboratories that not only test compliance with the standard but make practical interoperability tests as well.

2. Note that the topic of electronic purses is not considered in this working group and would have remained unsolved. A group of organizations lead by Europay, Visa, and the German Zentraler Kreditausschuß has therefore published the *Common Electronic Purse Specifications* (CEPS). The combination of EMV credit/debit features and a CEPS purse is possible and could be the best solution for the next decade.

reorganized *EMV 2000 Integrated Circuit Card Specification, Version 4.0*, had been issued. Version 4.0 corrects some EMV '96 errors that resulted in EMV-compliant cards not being compliant with the ISO/IEC 7816 standard. Version 4.0 also integrates specifications for the use of EMV cards for e-commerce transactions according to the Secure Electronic Transactions (SET) standard. The latest specifications can always be found at http://www.emvco.com/, which is the Web site of the EMVCo company founded in 1999 by Europay, MasterCard, and Visa to manage, maintain, and enhance the *EMV Integrated Circuit Card Specifications for Payment Systems*.

We should mention that EMV specifications define only those issues that are relevant for the interoperability of cards, terminals, and transaction processing throughout the world. They do not provide implementation guidelines or processing details that would concern no one but the issuers of the cards. Therefore, in addition to the EMV specifications, several more detailed specifications have been published with respect to the special requirements of the individual card organization. On the basis of the EMV '96 specification, there is the *Visa Integrated Circuit Card, Card Specification, Version 1.3.2* for Visa and the *MasterCard Chip—Recommended Specifications for Debit and Credit, Version 2.1* and the *Minimum Card Requirements for Issuance of Chip Pay Now (Debit) and Pay Later (Credit) Cards Version 2.1*, for MasterCard/Europay. Be aware that this is not a complete list of relevant specifications; you should check with the organizations before starting to develop a commercial EMV implementation. Newer versions now under development will be based on the EMV 2000 standard.

Because the MasterCard/Europay specifications leave more options open to the issuer, the definition of more restrictive card profiles is necessary. The first is the *Off-the-Shelf Card Profile*, which defines a not-too-complex subset of EMV in detail. Many smart card manufacturers have added smart cards with what are known as *M/CHIP Lite* operating systems to their product list. Although they all comply with the relevant Europay/MasterCard specifications, they may still differ in their initialization and personalization and additional functions. The same is true for cards compliant with the Visa specifications. EMV terminals have to be able to make transactions with both Europay/MasterCard- and Visa-compliant cards.

EMV cards and terminals have to be type approved by Europay/MasterCard or Visa, respectively. The two defined levels of type approval are level 1 and level 2. Level 1 defines mainly the mechanical and electrical characteristics as in ISO/IEC 7816-1, -2 and -3. Both protocols T = 0 and T = 1 are permitted. A level-2 approval (additionally) requires the correct functional behavior of the card according to the *EMV Integrated Circuit Card Specification* and the related specifications of either Europay/MasterCard or Visa. Only if a card and a terminal are level 2–type approved will they be interoperable.

Note that a terminal alone cannot really be level-2 compliant, but merely level-2 prevalidated. Full level-2 compliance can only be achieved in combination with the background system.

7.2 EMV Transactions

During an EMV payment transaction, a terminal has to perform several steps with the card. Figure 7.2 gives an overview of the transaction phases. At first the terminal selects an appropriate application (for example, a Visa credit application) on the card by use of the assigned AID. If more than one application is supported by the card and the terminal, then the cardholder must choose which one to use.

Application selection can be performed either explicitly or implicitly. Implicit selection means reading the contents of a standardized directory structure that shows all applications available on the card. If this structure is not available at the card, the terminal tries to select explicitly every application that it supports.

After application selection, the terminal reads all relevant card data. The command GET PROCESSING OPTIONS delivers the application file locator (AFL), which is a list of short file identifiers and record numbers that the terminal needs to read from the card, and the application interchange profile (AIP), which is a list of functions to be performed in processing a transaction with this card. Then the data such as cardholder name, primary account number (PAN), application expiration date, application version number, cardholder verification method (CVM), issuer country code, and the relevant public keys as well as their certificates are read with the READ RECORD command. The AFL concept allows the card manufacturer to choose which data to put in which records of which files while maintaining interoperability.

In the third step, the authenticity of the card is checked offline by either static or dynamic data authentication. Both processes are based on asymmetric cryptography with the RSA algorithm. Static data authentication means that a concatenation of fixed card data is signed with the issuer's private key. This signature is written to the card during the personalization phase and remains "static." The terminal can verify if the card data have been

Select | Application selection

Get processing options
Read record | Initiate application
Read application data

Internal authenticate | Off-line data authentication

Processing restrictions

Verify | Cardholder verification

Terminal action analysis

Generate AC | Card action analysis ----→ On-line Processing

External authenticate
Generate AC | Completion ←----

Application block
Application unblock
Card block
PIN change/unblock
Update record
Put data | Script processing

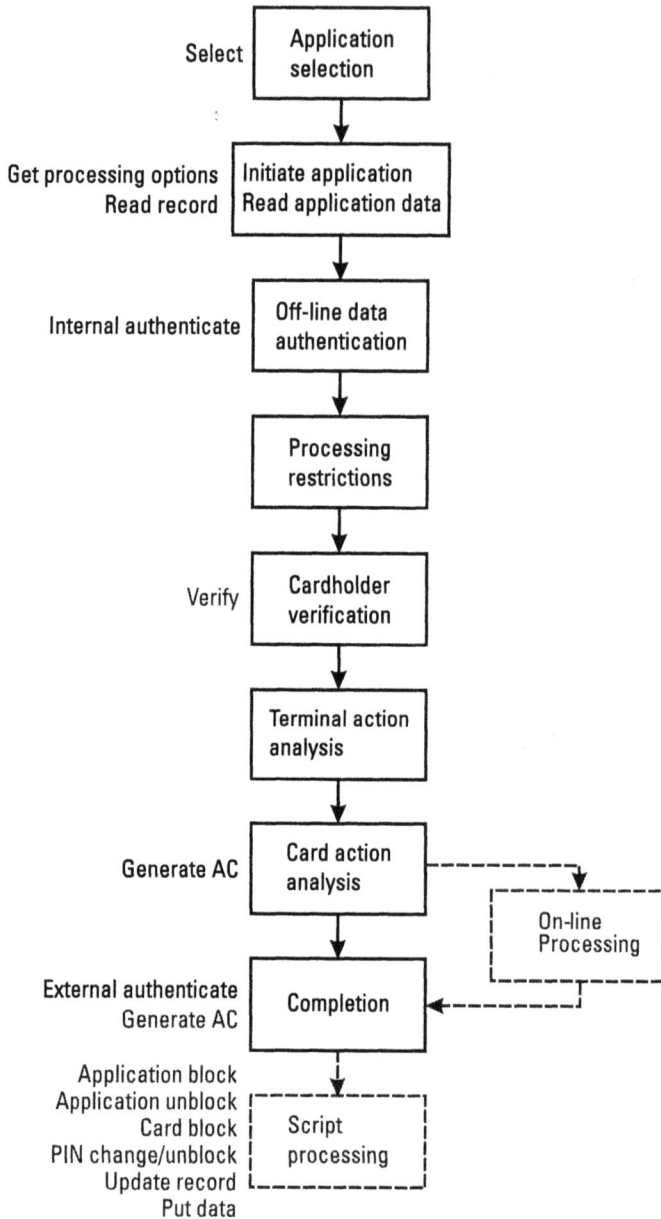

Figure 7.2 EMV transaction phases.

modified by checking the signature with the issuer's public key. The authenticity of the public key is verified by the certificate, which is signed with a

certification authority's key for which every terminal holds the corresponding public key.

We can see that static data authentication does not prevent duplicate cards, because all necessary data including the signature can be read out without restrictions. Hence, dynamic data authentication is preferred. In this case the card itself holds its own private/public key pair and a corresponding certificate signed with the issuer's public key (see Figure 7.3).

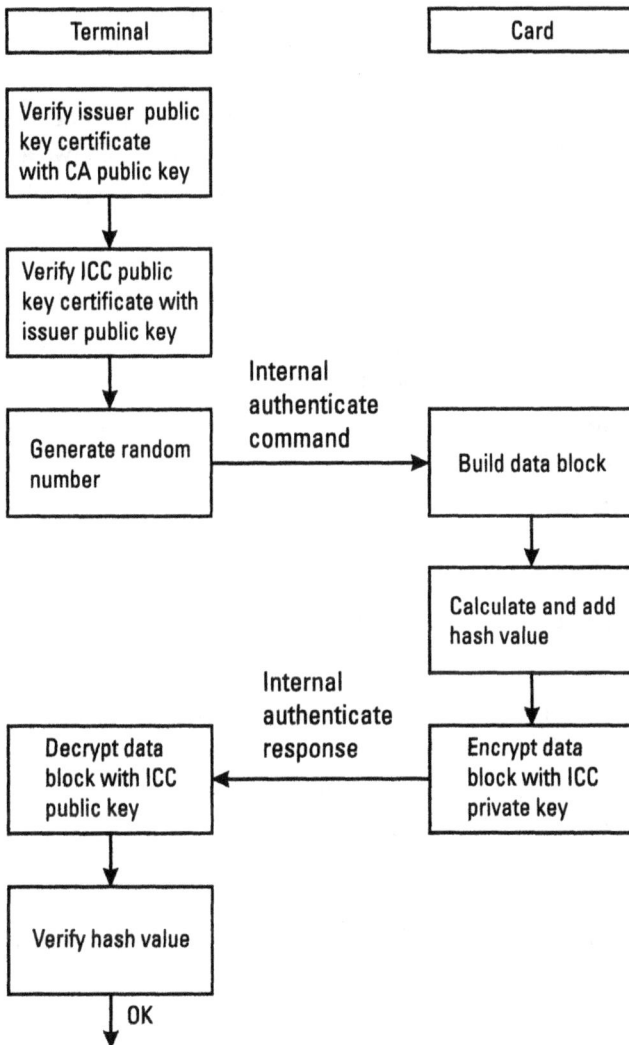

```
   ┌─────────────────┐              ┌─────────────────┐
   │    Terminal     │              │      Card       │
   └─────────────────┘              └─────────────────┘

┌─────────────────┐
│ Verify issuer public │
│ key certificate      │
│ with CA public key   │
└─────────────────┘
        │
        ▼
┌─────────────────┐
│ Verify ICC public   │
│ key certificate with │
│ issuer public key    │
└─────────────────┘
        │
        ▼
┌─────────────────┐   Internal        ┌─────────────────┐
│ Generate random │   authenticate    │ Build data block │
│ number          │──command────────▶ │                  │
└─────────────────┘                   └─────────────────┘
                                               │
                                               ▼
                                      ┌─────────────────┐
                                      │ Calculate and add │
                                      │ hash value        │
                                      └─────────────────┘
                                               │
                        Internal               ▼
┌─────────────────┐     authenticate  ┌─────────────────┐
│ Decrypt data    │     response      │ Encrypt data     │
│ block with ICC  │◀──────────────────│ block with ICC   │
│ public key      │                   │ private key      │
└─────────────────┘                   └─────────────────┘
        │
        ▼
┌─────────────────┐
│ Verify hash value │
└─────────────────┘
        │ OK
        ▼
```

Figure 7.3 Dynamic data authentication (DDA).

The terminal sends an INTERNAL AUTHENTICATE command where the challenge consists of at least one unpredictable number. The card concatenates this challenge to the ICC dynamic data, which can consist of fixed and changing data elements, for example, application PAN and a transaction counter value. Then a hash value is calculated and concatenated to the data block as well.

Next, the whole data block is encrypted with the card's private key. Finally, the card responds with the encrypted data block. The terminal checks the authenticity of the card's public key in two steps and uses it to decrypt the data block. Afterward, the correctness of the signature can be verified, which completes the dynamic data authentication.

The fourth task of the terminal is to evaluate the processing restrictions. The necessary card data have already been read in step two and are now compared to the related terminal data. Examples of checks performed in this transaction phase include the following:

- Does the terminal support the version of the card application?

- Is the terminal capable of all functions the card requests, for example, cardholder verification?

- Is the terminal's current date within the validity period of the card application?

- Does the card allow the requested type of the transaction (ATM, cash, cash back, buying goods, buying services) in the terminal's country?

- Is the card listed in a blacklist?

Cardholder verification is the fifth step in an EMV transaction. Depending on the issuer's CVM rules for domestic and international transactions of different amounts, which were read from the card in step two, the terminal has to execute one of the following procedures:

- No CVM (easy to implement);

- Online PIN verification (PIN encrypted with symmetric cryptography);

- Offline PIN verification (PIN not encrypted);

- Offline PIN verification (PIN encrypted with asymmetric cryptography);

- Handwritten signature;
- A combination of signature and some PIN verification.

In the future biometrics will also be supported. Offline PIN verifications are performed by the card by means of the VERIFY command, and the card keeps a PIN try counter and a PIN try limit. If the PIN try limit is exceeded, the card will not allow further PIN verifications.

The sixth step is named terminal action analysis and produces the terminal's decision about whether and how to perform the transaction. Three outcomes of this decision-making process are possible:

- Approve the transaction offline;
- Decline the transaction offline;
- Process the transaction online with the issuer.

The rules on which the decision is made are partially stored in the terminal and partially read off the card in step two. The latter are called issuer action codes (IACs). Here are some examples of rules:

- If the offline PIN verification fails, the terminal will decide to go online.
- If the merchant presses a certain button, the terminal will decide to go online.
- If the transaction amount is higher than a certain limit, the terminal will decide to go online.

The resulting decision is sent to the card via the GENERATE APPLICATION CRYPTOGRAM (AC) command. The terminal requests a transaction certificate (TC) to approve the transaction, an application authentication cryptogram (AAC) to decline the transaction, or an authorization request cryptogram (ARQC) to go online.

The card is not obliged to follow the decision of the terminal. First it performs its own decision-making process, called card action analysis. Even if the terminal has requested a TC, the card can decide otherwise and respond with an ARQC or an AAC. Of course, once a transaction is declined by the terminal, the card is not allowed to approve it. A request to go online to the issuer may in general also be declined, but in certain special cases the card is not allowed to refuse to go online.

The rules of the card are similar to those stated earlier for the terminal. Other possible rules include these:

- If the card is new and is making its first transaction, the card will decide to go online.

- The card will decide to go online after n consecutive offline transactions. This is called velocity checking.

- The card will decide to go online if the accumulated amount of all offline transactions since the last online transaction exceeds a specific limit. This is another type of velocity checking. Limits may differ for domestic and international transactions.

- If during the previous transaction an online issuer authentication was requested but not performed, the card will decide to go online again.

The response of the card to the GENERATE AC command shows the type of response (TC, AAC, or ARQC), the current transaction counter value, and the cryptogram. The 8-byte cryptogram is calculated by applying a DES-based cryptographic algorithm to a data block containing transaction type, amount, date, an unpredictable number from the terminal, the results of terminal and card action analysis, transaction counter, and similar data.

Step eight, called online processing, is only performed if requested by the card in response to the first GENERATE AC. Online processing not only allows the issuer to check the transaction before it is approved, but enables the card to authenticate the issuer by requesting a cryptogram, which may be presented to the card either in a separate EXTERNAL AUTHENTICATE command or included in the second GENERATE AC. Another option is the issuer script, which allows the issuer to send a number of different card commands to the terminal; typical examples are APPLICATION BLOCK, APPLICATION UNBLOCK, CARD BLOCK, PIN CHANGE/UNBLOCK, or the updating of card parameters with the commands UPDATE RECORD or PUT DATA. The terminal will carry out these commands on the card directly before or after completion of the transaction.

After the first GENERATE AC and any online processing, the transaction is completed with a second GENERATE AC. This is similar to the first one, but requests to go online (ARQC cryptograms) are no longer allowed. The card performs another card action analysis and thus can react to the result of any issuer authentication.

7.3   EMV 2000 Details

This section offers more details from the EMV 2000 specifications, which are subdivided into four books. That structure is adopted here.

7.3.1   EMV Book 1

Book 1 of the EMV 2000 specifications covers electromechanical characteristics, logical interface, and transmission protocols, as well as files, commands, and application selection.

The physical and electrical characteristics of the card are generally specified to be the same as in ISO/IEC 7816-1 and ISO/IEC 7816-2. Only a few details are more restrictive. For example, it is explicitly stated that no programming voltage shall be used. The only supply voltage allowed is 5± 0.5V, and the valid clock frequency range is from 1 to 5 MHz. One interesting note regarding the terminal says that the ICC should always be accessible to the cardholder. ATMs and other terminals that pull the whole card into themselves have to provide an eject mechanism that works even in the case of power loss.

The physical layer of communication follows ISO/IEC 7816-3 (see Chapter 5). Even proprietary protocols are allowed to be supported by the card, but in addition to such protocols one of the two standardized protocols $T = 0$ or $T = 1$ must be supported and be offered as first choice. The ATR indicates these protocols according to the ISO standard. Terminals must support both $T = 0$ and $T = 1$.

The second part of Book 1 covers files, commands, and application selection. The ISO dedicated files are subdivided into directory definition files (DDFs) which may contain other DFs, and application definition files (ADFs), which shall contain only application elementary files (AEFs). Every application on the card, for instance, a Visa credit beside a Maestro debit application (an unlikely case), resides in its own ADF and has an AID conforming to ISO/IEC 7816-5. The first 5 bytes of the AID denote the organization, for example, Europay, MasterCard, or Visa, and the next 11 bytes denote a certain function, for example, a Visa credit function or a Maestro debit function.

For easier application selection, there may be a payment system directory file containing a list of all supported applications. At the top level of the payment systems environment (PSE), there is a DDF with the standardized name "1PAY.SYS.DDF01." It contains an EF that is the payment system directory. The data from the EF can be read by use of the READ RECORD command. Each record gives the AID and the location of an application residing on the card.

The terminal can then compare this list of AIDs offered by the card with its own list of supported applications and may subsequently choose one application according to its internal priority list or offer the common list of applications for the customer's choice. If there is no payment system directory file on the card, the terminal will simply try to select all applications of its own list.

The two commands needed for application selection, SELECT and READ RECORD, are also specified in Book 1. The other card commands follow in Book 3.

7.3.2   EMV Book 2

Book 2 covers the issues of security and key management, in particular static and dynamic data authentication, PIN enciphering, application cryptograms, issuer authentication, secure messaging, and principles and policies for certification authorities and terminal security. The security architecture is built in such a way that the terminal does not hold any secret keys and hence does not need a security module (that is typically a built-in terminal card). This is achieved by predominant use of asymmetric cryptography.

The principles of static and dynamic data authentication have already been explained in Section 7.2. If there is more than one EMV application on the card, they may use the same cryptographic keys for the dynamic data authentication. This is important because RSA keys and certificates need significant memory space, and payment cards are usually equipped with ICs from the medium price range. Besides the standard dynamic data authentication as described in Section 7.2, an optional combined dynamic data authentication/application cryptogram generation is specified. Here the step of performing an INTERNAL AUTHENTICATION command is replaced by including the challenge/response authentication in the first GENERATE AC command.

PIN enciphering is also built on asymmetric cryptography in order to avoid the need for a terminal card. If the terminal is one secure device containing both the card interface and the PIN pad, there is no need to encipher the PIN. In practice, however, the terminal may be split into parts connected via a cable that is not at all tamper-proof. In this case PIN enciphering is required at the PIN pad or at a device securely connected to the PIN pad. For this purpose the card holds a separate key pair with a corresponding certificate signed with the issuer's private key. Before encrypting the PIN with the PIN enciphering public key, the terminal requests a random number by sending a GET CHALLENGE command. This random number is concatenated to the PIN before encryption to prevent replay attacks.

The GENERATE AC command uses symmetric cryptography (Triple-DES) because this step must be performed even on low-cost cards. The card derives a session key from a master key using the transaction counter. This session key is then used to calculate a MAC for certain transaction data such as transaction amount, terminal country code, terminal verification results, transaction currency code, transaction date, transaction type, application interchange profile, and a random number from the terminal.

Secure messaging works basically as in ISO/IEC 7816-4 (see Section 4.4).

7.3.3   EMV Book 3

Book 3 specifies the data elements, the files in the card, the commands, and the transaction flow. The latter has already been discussed in Section 7.2.

The tag-length-value (TLV)–coded data objects are stored in linear fixed or linear variable record AEFs. The short file identifiers 1 to 10 are reserved for EMV 2000 data; more payment system or issuer-specific data shall use the SFIs 11 to 20 and 21 to 30, respectively. In several cases the card asks the terminal for a list of data objects that is not TLV encoded. Because the content of any of these lists depends on the card, the card contains a data object list (DOL) specifying the expected order and format of the data objects. Examples are the optional processing options data object list (PODL), which lists the terminal-resident data elements needed by the card when carrying out the GET PROCESSING OPTIONS command, and the card risk management data object lists (CDOL1 and CDOL2) for the two GENERATE AC commands.

Those EMV card operating system commands that are not standardized by ISO/IEC have their CLA byte set to 8x H. The 9x H class is reserved for operating system manufacturers' proprietary commands, for example, commands needed during initialization and personalization of the IC. The ExH class is dedicated to issuers' proprietary commands. The following commands are specified in Book 3: APPLICATION

BLOCK and APPLICATION UNBLOCK, CARD BLOCK, EXTERNAL AUTHENTICATE, GENERATE AC, GET CHALLENGE, GET DATA, GET PROCESSING OPTIONS, INTERNAL AUTHENTICATE, PIN CHANGE/ UNBLOCK, READ RECORD, and VERIFY.

The transaction flow for what is known as *chip electronic commerce* is also included in Book 3. This part specifies the use of EMV cards in SET environment. SET was introduced some years ago but has not yet become widespread. Although even today's software-only SET transaction scheme provides much better security than simple Secure Sockets Layer (SSL) transmission of credit card data, the system seems too complex, and the customer's certificates are stored on the hard disk, which is uncomfortable in many respects. It is expected that EMV chip electronic commerce will reach a high level of adoption once millions of EMV cards have been distributed.

7.3.4   EMV Book 4

Book 4 describes the cardholder, attendant, and acquirer interface requirements, also specifying terminal types with their physical characteristics, functional requirements, and their software architecture. These issues are too far removed from the card to be of further interest here.

Part II
Java Card

Java Card is a platform for developing Java applications for smart cards. This part of the book deals with Java Card features, including its architecture, specifications, and manufacturer-specific aspects. Because Java Card can host multiple applications that do not necessarily trust each other, the Java Card security concepts are explained in detail. Part II ends with a general overview of the Java Card application development principles as an introduction to Part IV.

8

Java Card Basics

This chapter is dedicated to the basics of Java Card technology. It first explains the Java Card architecture and its remarkable features. Particular attention will be paid to differences between Java Card and Java technologies. The section concludes with the major steps of Java Card application development and explains the Java Card application programming interface (API).

8.1 Java Card Architecture

One of the main ideas that encouraged the development of Java Card technology was to make smart card applications portable across different platforms. The advantages of Java, such as platform independence and language-level security, were already well known and appreciated. Hence, a plan to bring the power of Java to the world of smart cards emerged and was implemented. The Java Card platform is formed by a combination of a customized subset of the Java programming language and a Java run-time environment dedicated to smart cards and other resource-constrained devices. Due to the fact that smart cards still have low performance and limited resources, it was necessary to customize the Java language, which was initially developed for the world of traditional computers. Customization of Java resulted in the omission of some features that are either impossible to implement on a smart card or irrelevant to smart card applications. However, some features were also added to Java Card to accommodate the specifics of a smart card and its applications. These are discussed in detail in later sections of this chapter.

Java Card is characterized by the following major benefits:

- *Platform independence.* Java Card applications written in accordance with the specifications are intended to run on any Java Card–compliant smart card. This feature was thought to ensure a high degree of portability of Java Card applications. Unfortunately, individual smart card manufacturers frequently introduce their own packages with a manufacturer-dependent API (especially security-related APIs) or still support different versions of Java Card. This significantly decreases the portability of Java Card applications.

- *Multiple-application support.* More than one application can be run on a Java Card technology smart card. Furthermore, the data of each application is securely protected from any other application run on the same card.

- *Power of Java.* Java Card inherits many benefits of the Java programming language. In the particular case of smart cards, such benefits are object-oriented programming and language-level security. However, some limitations on Java introduced in Java Card (see Section 8.2) frequently lead to a style of programming that is different from conventional Java. Another advantage of Java Card is that its applications can be developed using any development tool or environment for standard Java.

The Java Card architecture is illustrated in Figure 8.1. As can be seen, it looks very similar to traditional Java. The smart card operating system (OS) is layered on top of a smart card microcontroller and is aimed at providing common services like file and data management, communication, and command execution. From the communication point of view, Java Card is fully compliant with ISO/IEC 7816. In particular, Java Card supports communication protocols[1] and commands in accordance with ISO/IEC 7816-3 and ISO/IEC 7816-4, respectively.

The Java Card run-time environment (JCRE) is layered on top of the smart card operating system and consists of the Java Card Virtual Machine (JCVM), the Java Card API, also referred to as the framework, and native

---

1. T = 0 and T = 1 protocols.

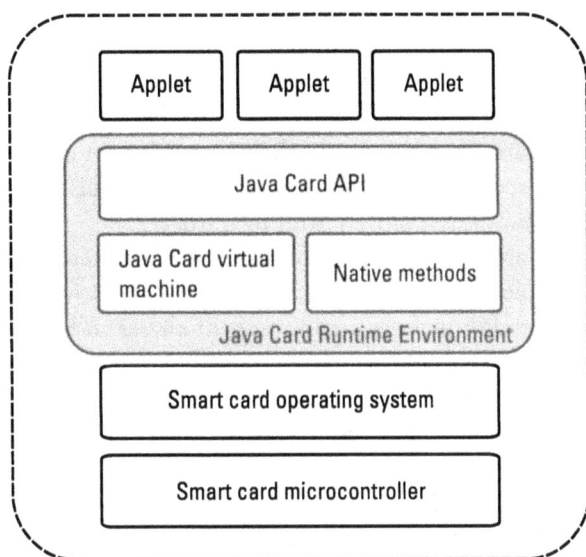

Figure 8.1 Java Card architecture.

methods. Native methods are needed to implement certain special platform-dependent operations like I/O operations or cryptographic operations in a compact and efficient way. That is why the implementation of such operations interacts directly with the smart card operating system and is usually done in languages other than Java (typically, C or Assembler). The Java Card API is formed by a number of packages containing classes dedicated to various purposes (see Chapter 11). In addition to the standard Java Card API, particular JCRE implementations frequently contain some manufacturer-specific extension APIs. On the one hand, they provide some additional functions, but on the other, they decrease the cross-platform portability of Java Card applications.

Java Card applications, called *card applets* or simply *applets,* written in the Java programming language are located on the topmost level of the Java Card architecture. More than one applet can be run on a card. Each applet on a card is uniquely identified by its AID. Chapter 10 of this book addresses security issues involved with the Java Card's multiple-application support.

The main task of the JCVM is to execute an applet bytecode on a card and to provide the Java language support. The core difference between the JCVM and the conventional Java Virtual Machine is that the first one is actually split into two independent parts. One part of JCVM, called the Java Card Converter, is executed off-card, for instance, on a personal computer.

The second part of JCVM is run on-card and is capable of applet code execution, managing classes, and providing interapplet security mechanisms. In contrast to Java, the lifetime of the on-card JCVM is limited only by the lifetime of a smart card. In other words, the on-card JCVM cannot be stopped and then started new again—it always runs on a card and is merely temporarily paused when power is removed from the card.

The Java Card Converter is a software tool that prepares a card applet bytecode (all applet `class` files put into one package) for uploading to a card. This preparation includes verification of classes to be loaded, various checks for Java Card–specific restrictions and violations, allocation and creation of the applet data structures, and resolution of symbolic references to the applet data structures. The result of the conversion is a converted applet (`cap`) file containing a complete image of the applet prepared and optimized for an execution on a card.

Figure 8.2 illustrates the principle described above and shows the main steps of card applet development. A card applet code can be written and compiled using any Java development tool and environment. Debugging and testing is a different case—because of the specifics of Java Card and the use of manufacturer-specific packages, this can be done in most cases only with the help of development tools provided by the smart card manufacturer.

After compilation of all source `java` files related to the applet, the resulting class files are passed to the Java Card Converter, which generates the applet `cap` file as an output. The applet `cap` file then can be uploaded to a card. Java Card specifications do not define exactly how the applet `cap` file is uploaded to a card—this also remains a manufacturer-specific issue.

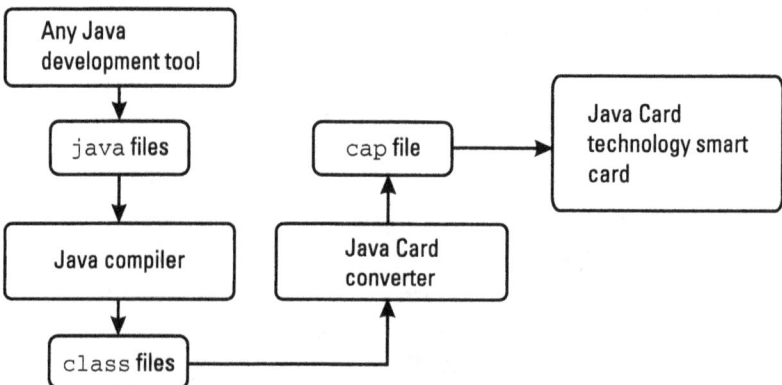

Figure 8.2 Java Card applet preparation.

A few words must be said about how an applet is uploaded to a card. Any Java Card technology smart card contains a special application called the installation program that is capable of loading an applet cap file and storing it on the card. Thus, there is no need for JCVM to take care of loading the applet—this is accomplished by the installation program. From an architectural point of view, the installation program can be seen as an ordinary Java Card application layered on top of JCVM and implementing an applet cap file upload over certain format APDUs sent to the card by a terminal.

Note that, in order to increase applet uploading security, certain JCRE implementations allow the applet cap file to be uploaded in a digitally signed and encrypted manner. In this case, the applet is uploaded successfully only if the applet's digital signature is successfully verified on the card.

Another remarkable feature of Java Card is that it does not provide ISO/IEC 7816-4 file system support on-card. In other words, the Java Card API has no means of working with files in terms of creating, writing, reading, and so forth. All functions related to file representation and handling should be implemented within an applet. Although this looks like a restriction, it gives more flexibility and allows implementation of only those file support features that are really needed by an applet.

Initially, the plan was to provide file system support on Java Card. Even the previous version of Java Card, Java Card 2.0, contained a set of classes dedicated to operations on files. It is said that manufacturers could not come to an agreement on an underlying API and therefore file system support was left out of Java Card 2.1.

A practical object-oriented implementation of a Java Card file system is demonstrated in Part IV of this book.

JCVM, JCRE, and the Java Card API are defined by Sun Microsystems Inc. specifications [1–3], which are available online.[2] As of February 2001, not all existing Java Card implementations were based on Java Card 2.1. For instance, iButton from Dallas Semiconductor and Schlumberger Cyberflex follow the Java Card 2.0 specification.

In May 2000, the Java Card 2.1.1 specification was released [4]. In comparison with Java Card 2.1, Java Card 2.1.1 contains a number of minor improvements and pays more attention to some aspects of Java Card implementation.

---

2. http://java.sun.com/ products/javacard.

8.2 Differences from Java

A smart card is a resource-constrained device. It cannot provide the amounts of memory and high performance that are available on modern computer architectures. That is why it is impossible to implement the standard Java platform in a one-to-one manner on a smart card. The decision was made, therefore, to implement Java Card as a subset of standard Java, omitting some features and adding some restrictions.

First of all, because of the resource constraints and limited CPU performance, Java Card does not support multithreading. Second, Java Card does not support dynamic class loading, for an obvious reason: It is very problematic and almost impossible to ensure loading of additional classes to the card during applet execution. Object cloning is also not supported by Java Card.

All objects once created by an applet will exist as long as the applet exists, that is, until the applet is deleted from the card. This means that all objects[3] created by the applet are persistent, that is, their values are preserved when power is removed from the card. Therefore, Java Card does not need and does not support garbage collection. As a consequence, the method finalize() is not supported. This feature also increases applet safety: References to nonexistent objects are avoided because objects cannot be destroyed during an applet's lifetime. On the other hand, implementation of garbage collection could be quite useful in that it could prevent a loss of memory occupied by a dynamic object that leaves the applet's scope. Some Java Card implementations, like iButton from Dallas Semiconductors, support garbage collection.

The following sections discuss in detail certain differences between Java Card and Java.

8.2.1 Primitive Data Types and Arrays

Like Java, Java Card supports such primitive data types as byte, short, and boolean. A byte is an 8-bit signed number with values that can range from −128 to 127. A short is a 16-bit signed number with values that can range from −32,768 to 32,767. A boolean value is represented internally by a byte.

In contrast to Java, Java Card does not support such data types as float, double, long, and char at all. Data type int is optional; that is, some particular Java Card implementations may support it, some not. A summary of supported and unsupported Java Card primitive data types is given in Table 8.1.

---

3. Except transient objects that are created in a special manner and whose value is reset upon certain Java Card system events.

Table 8.1
Supported and Unsupported Primitive Data Types in Java Card

Data Type	Width (bits)	Supported?
byte	8	Yes
short	16	Yes
boolean	8	Yes
int	32	Optional
char	16	No
float	32	No
long	64	No
double	64	No

Java Card supports only one-dimensional arrays, not multidimensional arrays. This limitation is also because of the limited resources available on a Java Card technology smart card. As in Java, elements of an array may be of any supported primitive data type or objects. The following example demonstrates valid declarations of arrays:

```
byte byte_array[] = new byte[3];
byte states[] = {0, 1, 2} ;
PIN app_pins[] = new PIN[3];      // array containing 3
// references to PIN objects
```

The following array declarations are invalid because they declare multidimensional arrays:

```
byte a[][] = new byte[3][3];
boolean flags[][] = new boolean[5][5];
```

As in Java, Java Card arrays are represented by objects. This means that methods of the class Object can be applied to them. For instance, an equality of two array references can be checked using the method equals() of the Object class:

```
if ( states.equals(byte_array) ) {
    ...
}
```

The method returns a boolean value indicating whether the array references are equal or not. More advanced operations on arrays (copying, comparing, etc.) can be performed with the help of static methods of the class `Util`, which is a member of the Java Card framework classes.

8.2.2 Operations and Type Casting

Java Card supports all arithmetic, logical, and bit-wise operations defined in Java. However, typecasting rules used in Java Card are slightly different from rules defined in Java. The main typecasting rule of Java Card states that results of intermediate or unassigned operations must be explicitly cast to a type of a desired value. An intermediate calculation is part of a complex expression involving a number of operations on a number of values. A result of an unassigned operation is not assigned to any variable. An example of an unassigned operation could be an array index calculation.

The reason behind the explicit typecasting rule is that, in Java, results of intermediate or unassigned operations are cast to the type `int` by default. However, Java Card supports the type `int` only optionally, which implies that not all Java Card implementations will have it. Hence, casting either to the types `short` or `byte` must be specified explicitly. The following example demonstrates correct explicit casting of results of intermediate or unassigned operations:

```
byte byte_array[] = new byte[3];
byte b;
short s;
b = byte_array[(byte)(s-1)]; // unassigned operation
b = (byte)( (byte)(s+6)*2 ); // intermediate operation
```

The example below demonstrates erroneous typecasting:

```
b = byte_array[s+1];
b = (byte)( (s+6)*2 );
```

Typecasting errors related to Java Card restrictions are reported by the Java Card Converter.

8.2.3 Exceptions

In principle, Java Card supports all Java mechanisms for exception handling. Card applets may contain `try`, `catch`, and `finally` statements.

Obviously, exceptions related to unsupported features, like multithreading or dynamic class loading, are not supported. Moreover, the constrained resources of a smart card also have an impact, resulting in the following three features of Java Card exception handling:

1. Not all of the Java exception classes are supported.

2. Descriptive string messages in exceptions are not supported. Instead, a reason code of the type `short` is used.

3. Creating instances of exception classes is not recommended. Instead, static JCRE instances of exception classes should be used.

We now discuss each aspect of this list in detail. All Java Card exceptions are subclasses of a superclass `Throwable`. Exception classes are stored in two core packages of the Java Card framework, `java.lang` and `javacard.framework`. Exceptions contained in the first package represent erroneous situations related to Java language programming. Table 8.2 gives a general overview of exceptions contained in the `java.lang` package.

Table 8.2
`java.lang` Package Exceptions

Exception	Description
ArithmeticException	Indicates a certain arithmetic run-time error. An example could be the division-by-zero error.
ArrayIndexOutOfBoundsException	Indicates that an array index is outside of the array boundaries.
ArrayStoreException	Indicates that there was an attempt to store an object of an incorrect type in an array.
ClassCastException	Indicates an incorrect attempt to cast an instance of one class to another class.
NegativeArraySizeException	Indicates an attempt to create an array with a negative size.
NullPointerException	Indicates a null reference access.
SecurityException	Indicates a violation of access rights for a certain object.

One important fact must be mentioned: Java Card specifications do not define JCVM behavior for the case in which a certain exception is thrown and is not caught by a card applet. As a first consequence of an uncaught exception, JCVM will halt, that is, card applet execution will be stopped. What will happen then depends on the particular Java Card implementation. For instance, the Sm@rtCafé Java Card technology smart card from Giesecke & Devrient, which is used to implement a sample EMV application later in this book, will respond to a terminal with a status word indicating a general card error.

Exceptions contained in the `javacard.framework` package represent smart card–specific erroneous situations that occur during a card applet execution. Table 8.3 gives their general description.

Java Card does not support the object type `string`. Therefore, Java Card exceptions do not provide descriptive string messages. Instead, additional information about the reason for an exception is reported by a *reason code*. The reason code is a value of the type `short`. A remarkable thing about exception reason codes is that most exception classes, mainly smart card–specific exception classes, contain predefined static constants representing main reason codes typical of the underlying exception.

To conclude the description of Java Card exceptions, a few words must be said about exception usage. First of all, it is strongly recommended not to create a new exception object each time an exception is thrown. Instead, all exception objects needed by an applet should be created during the applet initialization phase, the references to them stored, and the objects reused

Table 8.3
`javacard.framework` Package Exceptions

Exception	Description
APDUException	Indicates errors related to APDU handling.
ISOException	Is used to issue a response APDU with a given status word.
PINException	Indicates errors related to PIN handling.
SystemException	Indicates errors occurring on the Java Card at system level.
TransactionException	Indicates errors occurring during transaction processing.
UserException	Is used to implement user-defined exceptions.

each time an exception must be thrown. In this context, "reused" means that an exception object is created just once but thrown as many times as needed with a desired reason code. The reasoning behind such a practice is obvious: Creating new instances of exception classes will simply waste the limited card memory available.

There is an even more efficient method of exception throwing. JCRE precreates all exceptions defined in the Java Card API. In other words, JCRE creates instances of all Java Card exceptions by default. This means that these precreated exception objects can be used instead of objects created by a card applet, so there is no need to create most of the exception objects at all. All exceptions defined in the `javacard.framework` package (see Table 8.3) have a static method `throwIt()` that throws a JCRE (a precreated) instance of the class.

Let us demonstrate this principle with an example. Assume that an applet must report that the instruction (INS value) given in a command APDU is not supported. This can be achieved with the following statement:

```
ISOException.throwIt(ISO7816.SW_INS_NOT_SUPPORTED);
```

First of all, execution of this statement will throw a JCRE instance of the `ISOException` exception class with the desired reason code. As a consequence, this exception will force JCRE to issue a response APDU with the ISO 7816-4 status word 6D 00 H defined by the static constant `SW_INS_NOT_SUPPORTED` of the Java Card framework interface `ISO7816`.

8.3 Java Card Applet

The lifetime of a Java Card applet consists of a number of stages. After being compiled and converted to a `cap` file (see Section 8.1) by the Java Card Converter, the applet is loaded to a card by the card installation program. This is the moment when the on-card life of the applet begins. First of all, the applet must be installed and registered within JCRE. If the applet registration is accomplished successfully, the applet becomes available for selection via SELECT APDU, sent to the card, and processed by JCRE. The selected applet is ready to receive incoming command APDUs delivered to it by JCRE, to process them, and to generate response APDUs that are sent out by JCRE.

As pointed out in Section 8.1, the lifetime of a card applet is limited by the lifetime of the Java Card Virtual Machine, that is, by the lifetime of the card. However, note that certain Java Card implementations may allow clearing of the application area of a card's EEPROM. In this way, all applets existing on the card and all data objects belonging to them are completely deleted from the card.

The Java Card API provides handy mechanisms for card applet implementation. Any card applet is implemented on the basis of an abstract base class `Applet` located in the `javacard.framework` package. The class `Applet` contains all methods necessary for applet installation, selection, and deselection, and APDU processing. Those methods and aspects related to them are discussed in detail in the following section.

8.3.1   Installation and Registration

After an applet has been successfully loaded to the card, it must be installed. The installation procedure is initiated by the INSTALL APDU sent to the card. Java Card specifications do not define the exact format of this APDU; they instead leave it up to the manufacturer. The INSTALL APDU is received and processed by the same card application that loaded the applet cap file to the card—the installation program.

On receiving the INSTALL APDU, the card installation program simply invokes a special method of the applet that is to be installed. This method is called `install` and is defined in the abstract class `Applet` extended by any card applet. The installation program also passes to the `install` method applet initialization options received with the INSTALL APDU. The applet `install` method is called only once (obviously, an applet is installed on a card only once).

The core task of the `install` method is to create an instance of the loaded applet class and to register the instance within JCRE. Naturally, the applet constructor is called when the applet instance is created. The constructor may create data objects used by the applet, and it is good programming practice to create all applet objects in the applet constructor.

The applet instance registration is mandatory: If it is not performed, the applet installation fails. The registration is done via invocation of the `register` method of the applet. The `register` method exists in two versions, one with parameters, the other without. The `register` method with parameters is used to specify an AID of the applet instance.

Summarizing everything said above, the main steps of an applet installation procedure (assuming that the applet is already loaded to the card) are as follows:

1. Card installation program receives INSTALL APDU and invokes the `install` method of the applet to be installed.
2. An instance of the applet class is created in the `install` method.
3. The applet instance is registered via invocation of the `register` method.

If the applet is installed successfully, JCRE makes it available for selection.

8.3.2   Selection and Deselection

Any applet installed on a card must be explicitly selected before command APDUs are sent to it. An applet is selected by means of the SELECT APDU with the following defined format:

CLA	INS	P1	P2	L$_c$	Data
00	A4	04	00	AID length	AID

The data field of the APDU contains an AID of the applet to select. Other fields of the SELECT APDU are fixed and defined in accordance with ISO/IEC 7816-4. If JCRE finds an applet with the given AID, it marks it as selected and forwards it to it all further command APDUs. If no applet with such an AID is found, JCRE reports the fact with the respective status word in the response APDU.

After a card reset, all applets on the card are in a suspended state. In other words, none of the applets is marked as selected. Therefore, if JCRE receives any[4] APDU different from SELECT, it will answer with the response APDU indicating that no applet is selected (status word 69 99 H). Note that some Java Card implementations may allow specification of a default applet. A default applet is marked as selected after a card reset and JCRE will forward to it all received APDUs even if there was no explicit SELECT command. However, Java Card 2.1 specifications address no means for defining a default applet and leave this question up to the manufacturer.

The abstract class `Applet` contains two methods related to applet selection and deselection. The first one is called `select()` and is invoked

---

4. Except manufacturer-proprietary command APDUs related to card personalization and management, for example, applet load or install APDUs. Command APDUs of this kind are not considered further in this discussion.

by JCRE whenever the applet becomes selected. An applet may perform operations needed for further processing of commands; for example, it may change the values of internal flags. The `select()` method should return a boolean value indicating whether it is ready to accept commands or not. By default, the value `true` is returned.

The applet method `deselect()` is called by JCRE when a currently selected applet becomes deselected, that is, when another applet on the card is selected. Obviously, this method is not called when power is removed from the card.

An interesting feature of SELECT APDU processing is that the APDU is also passed to the applet after its selection by JCRE. This means that the applet also has possibilities of processing this APDU and answering it in a desired manner.

Aspects related to the processing of command APDUs by an applet are addressed in the next section.

8.3.3   APDU Processing

Figure 8.3 demonstrates a general scheme for incoming APDU processing by JCRE. Applet selection mechanisms were presented in the previous section. The abstract class `Applet` extended by any Java Card applet contains the method `process`. This method is invoked by JCRE for each received command APDU. All operations dealing with processing the APDU, performing all necessary application-specific operations in response to the APDU, and preparing the response APDU are done in the applet `process` method.

The `process` method has one single parameter. This parameter is an instance of the `APDU` class, another Java Card framework class located in the `javacard.framework` package. This class provides a handy interface to the communication facilities of a smart card and is designed in a protocol-independent manner. Therefore, an applet developer does not have to deal with specifics of T = 0 or T = 1 protocols (those are the only protocols supported by Java Card 2.1)—all of them are "hidden" inside the `APDU` class and its methods implementation.

A core field of the `APDU` class is a byte array buffer that is used for reading data of the incoming APDU and preparing data of the outgoing (response) APDU. In addition, the class `APDU` provides a number of methods for easy access to the byte buffer.

If no exception is thrown during the `process` method execution, JCRE sends out data in the APDU buffer (if the response was constructed by the applet) with the success status word 90 00 H automatically attached. If

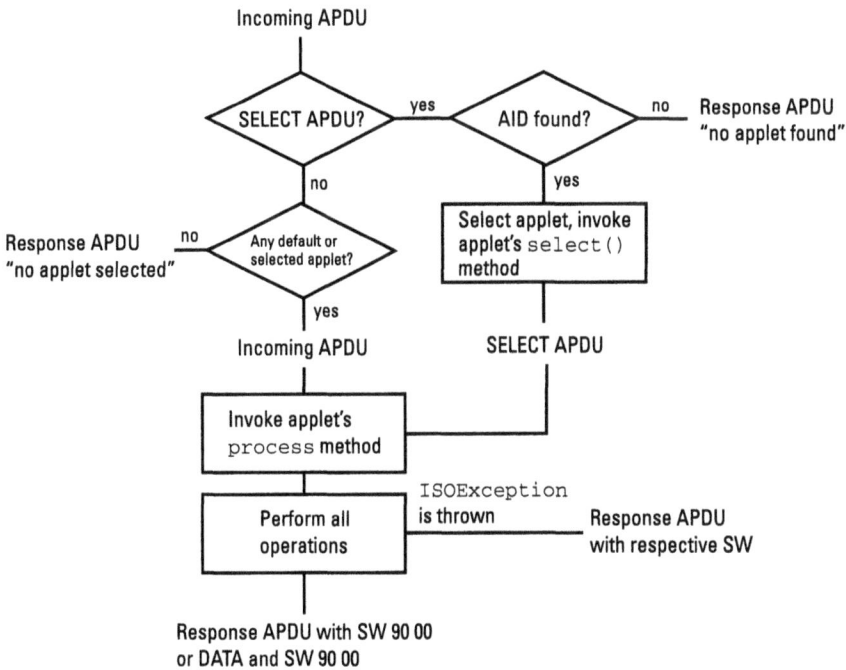

Figure 8.3 Command APDU processing by JCRE and an applet.

the applet throws an ISOException (see Section 8.2.3), JCRE catches it
and sends out a response APDU with the status word given in the exception
reason code. If any other exception is thrown during the process method
execution, JCRE will send out a response APDU with the status word "No
precise diagnosis" 6F 00 H.

The APDU class and the Applet calls are discussed in Chapter 11.

References

[1]   Sun Microsystems Inc., "Java Card 2.1 Virtual Machine Specification," Mar. 1999.

[2]   Sun Microsystems Inc., "Java Card 2.1 Runtime Environment (JCRE) Specification,"
       Feb. 1999.

[3]   Sun Microsystems Inc., "Java Card 2.1 Application Programming Interface," Feb.
       1999.

[4]   Sun Microsystems Inc., "Java Card 2.1.1 Specifications. Release Notes," May 2000.

9

Deployment of Java Card Technology

Smart cards can in general be used in mobile phones, personal digital assistants, set-top boxes, and other devices. Java Card technology supports platform independence, makes it possible to implement multiple applications on a single card in a secure way, and allows downloading of applications after a card has been issued. In addition, Java is a programming language in widespread use, which reduces the time-to-market for new smart card applications. All of these properties make Java Card interesting for a range of commercial applications. The following sections give some examples of Java Card technology deployment.

In addition to industry and financial institutions, government agencies have expressed strong interest in Java Card–based products. For example, in 1999, Citibank[1] issued multiple-application smart cards based on Java Card technology to General Services Administration employees. The cards provided a number of functions, including logical access, physical access, property management, e-ticketing, and e-boarding.

9.1 Java Card Forum

The Java Card Forum (JCF)[2] is an interindustry initiative to promote the Java Card API specification as the industry standard. It was founded by

---

1. http://www.citibank.org.
2. http://www.javacardforum.org.

Schlumberger and Gemplus in 1997 following JavaSoft's (a division by Sun Microsystems) announcement of the Java Card API in 1996. The member list includes chip manufacturers, card manufacturers, companies, and agencies in the financial, telecommunications, health care, transportation, and information technology sectors. Current work is focused on vertical market extensions to the core specification for GSM, banking, and information technology.

9.2   Card Management

As soon as one starts loading and unloading applications to and from smart cards after they have been issued, the problem of managing a card population arises. This is called the card management problem.

The Java Card Management (JCM) Task Force of the JCF was initiated in 1998 with the goal of defining a framework for a card management system (CMS). Specifically, the idea was to define a core CMS on top of which companies could build their own CMS, and to define on-card APIs for Java Card management to be used through an off-card CMS. For example, it is necessary to define the following core features:

- A card repository describing cards with general attributes;

- A card application repository describing the applets with general attributes;

- Card state management functions;

- Life cycle transition management functions;

- Post-issuance applet management.

As of 2001, one package is specified, org.javacardforum.management [1]. In addition to this document, several commercial card and application management specifications are available, such as MXI by MAOSCO,[3] Open Platform by Global Platform (see Section 9.4), and the Platform Management Architecture (PMA) by platform7.[4]

---

3. http:// www.multos.com.
4. http:// www.platform7.com.

9.3 SIM Application Toolkit

Mobile subscriber–relevant data and security algorithms are stored on the SIM (GSM 11.11 [2]).[5] The SIM can be implemented in two forms, either as a smart card or as a plug-in SIM. The SIM card initially played a "passive" role, providing the user with the authentication necessary to access the network and encryption keys to achieve speech confidentiality. SIM Application Toolkit, a part of the GSM standard (GSM 11.14 [3]), extends the card's role such that it becomes the interface between the mobile device and the network. SIM Toolkit supports the development of smart card applications for GSM networks. It is based on the client-server principle, with SMS as the bearer service. In the future, other transport mechanisms such as USSD or GPRS will be used. With SIM Toolkit it is possible to personalize a SIM card, to update existing SIM functions and services, and to install new functions and services by downloading data over the network. This has usually been done by adding or modifying data in the card files and records, not by downloading executable code.

In November 1999, ETSI adopted Java Card technology for inclusion in SIM Toolkit [4]. In the same year, ETSI issued a standard (GSM 03.19 [5]) describing the following extensions to the Java Card 2.1 API:

- The `sim.access` package provides the means for the applets to access the GSM data and file system of the GSM application defined in the GSM 11.11 specification.

- The `sim.toolkit` package provides the means for the toolkit applets to register the events of the toolkit framework, to handle TLV (tag-length-value) information, and to send proactive commands according to the GSM 11.14 specification.

The resulting cards provide GSM operators with the ability to deploy a wide range of value-added services, such as secure remote banking, stock trading, and unique dial-back roaming services. There are already Java Card 2.0–based SIM cards on the market, such as Giesecke & Devrient's StarSIM, Gemplus's GemXplore98, or Schlumberger's SIMera. Card applets can usually be transported to the card by SMS, either from a content provider or at a point-of-sale terminal. Cards have a Java Virtual Machine that supports the

---

5. GSM standards are issued by the European Telecommunications Standards Institute (ETSI); see http://www.etsi.org/.

sandbox security model, strong bytecode verification, and firewalls between card applets.

9.4 Visa Open Platform

An interesting development in the smart card and e-commerce area is the Visa Open Platform [6] supported by various financial institutions, service providers, mobile network operators, and hardware manufacturers. The goals are to develop standardized solutions for secure mobile electronic commerce and also an Open Platform chip that will allow financial institutions to dynamically download Visa payment applications to a mobile phone on the basis of Java Card technology. The technology is chosen in such a way that it ensures these goals will be reached:

- Interoperability of cards, terminals, operating systems, software products, and bank office support systems from different vendors;

- Secure support of multiple applications coexisting on the card (Java, Java Card, Windows for Smart Cards) in such a way that each application provider is assigned a separate security domain;

- Strongest commercially feasible security, which will be evaluated using the Common Criteria (see Chapter 3);

- Support of existing standards such as EMV (see Section 7.1) and ISO 7816 (see Section 1.5) so that the card can be used in the existing ISO/EMV-compliant terminals.

An Open Platform card could serve as a corporate credit card, stored-value purse for small purchases, security token for Internet commerce, or an electronic ticket carrier. Two specifications are relevant to smart cards:

1. The Open Platform Card Specification specifies the off-card communication with the terminal and the on-card communication with the applications. In other words, it defines the Open Platform API and how to use it to develop card applications.

2. The Visa Open Platform Card Specification defines the enhancements to the Open Platform that are needed to implement Visa-specific applications (e.g., cryptography support).

In the run-time environment, two different stacks are possible. One stack includes the Java Card Virtual Machine with the Java Card API layered over a proprietary card vendor operating system. Another stack includes Windows for Smart Cards (WfSC) with the corresponding Virtual Machine and API, layered over the WfSC operating system. In addition, the stack includes the Open Platform API, which extends the standard card API to allow additional security control (e.g., secure channel establishment, key verification before loading it on the card, card lock if a security threat is detected). The Open Platform environment also places some additional constraints on applications (e.g., secure card auditing, application loading after a card has been issued). For example, if an application is loaded after the card is issued, a secure channel is established between the card and the platform from which the application is loaded. In this way the card can authenticate the application provider, and the integrity of the application is guaranteed. The Open Platform defines its own card management (see also Section 9.1).

References

[1]   Java Card Management Task Force, "Java Card Management Specification," Version 1.0b, Oct. 2000; available at http://www.javacardforum.org/Documents/documents.html.

[2]   European Telecommunications Standards Institute, "Digital Cellular Telecommunications System (Phase 2+); Specification of the Subscriber Identity Module—Mobile Equipment (SIM-ME) Interface (GSM 11.11, Version 8.3.0 Release 1999)," 2000.

[3]   European Telecommunications Standards Institute, "Digital Cellular Telecommunications System (Phase 2+); Specification of the SIM Application Toolkit for the Subscriber Identity Module—Mobile Equipment (SIM-ME) Interface (GSM 11.14, Version 8.3.0 Release 1999)," Aug. 2000.

[4]   Hassler, V., *Security Fundamentals for E-Commerce*, Norwood, MA: Artech House, 2001.

[5]   European Telecommunications Standards Institute, "Digital Cellular Telecommunications System (Phase 2+); Subscriber Identity Module Application Programming Interface (SIM API); SIM API for Java Card (TM); Stage 2 (GSM 03.19, Version 7.1.0 Release 1998)," May 2000.

[6]   Visa International, "Visa Open Platform: Overview," 2000; available at http://www.visa.com/nt/suppliers/open/overview.html.

10

Java Card Security

Smart card security issues can be divided into four areas: (1) card body security, (2) hardware (i.e., chip) security, (3) operating system security, and (4) card application security. In addition to these general issues, which are addressed in Part I of this book, Java Card security encompasses the following areas: Java Card language subset security, card applet security mechanisms, and Java Card crypto APIs for writing secure programs. This chapter gives a brief overview of these aspects of Java Card.[1] For more details, please refer to [1].

10.1   Java Card Language Subset Security

As of mid-2001, Java was probably the most popular programming language [2, 3]. Its development started in 1991 at Sun Microsystems when James Gosling developed the Oak programming language. Oak was designed for consumer electronics software that could be downloaded (i.e., upgraded) over a network. The programs written in Oak were supposed to be very compact and highly reliable. Because portability (i.e., platform independence) was one of the major design goals, the source code was compiled into an interpreted byte-code to run on a virtual machine. In other words, the Oak bytecode contained a set of instructions not typical of any particular microprocessor, but for a specially designed "virtual microprocessor" (virtual machine).

---

1. http://java.sun.com/ products/javacard.

Java is a general-purpose object-oriented programming language similar to C++. It began from a subset of C++ in which all features considered error prone or unsafe were eliminated [4]. Some of Java's object-oriented properties are dynamic binding, garbage collection, and inheritance. Java programs are compiled into a processor-independent bytecode, which is loaded into a computer's memory by the Java Class Loader to be run on a Java Virtual Machine (JVM). JVM can run programs directly on an operating system or be embedded inside a Web browser. It can execute the Java bytecode directly by means of an interpreter, or use a "just-in-time" (JIT) compiler to convert the bytecode into the native machine code of the particular computer. JVM enforces Java safety, privacy, and isolation rules. These make it possible to protect against unauthorized access and to isolate one application from another within the same address space, so that it is not necessary to enforce address space separation between applications [5].

As a subset of the Java programming language and virtual machine specifications, the Java Card platform inherits the main Java security features such as Java safety and Java type safety, which are briefly described in the following two sections.

10.1.1 Java Safety

The term *safety* denotes the absence of undesirable behavior that can cause system hazards. Java is a safe programming language: Many of the confusing or poorly understood features of C++ cannot be found in it. For example, Java manages memory by reference and does not allow pointer arithmetic. Another feature that makes Java simpler and thus safer is that it does not allow multiple class inheritance. On the other hand, Java allows multiple interface inheritance. However, an interface, in contrast to classes, may not be used to define an abstract data type, since it may contain only constants and method declarations, and no implementations. Java also provides the final modifier, which disables subclassing when applied to class definitions and disables overriding when applied to method definitions.

In addition, some new mechanisms that can be programmed in C++ only by very experienced programmers are a part of the language in Java. For example, a useful mechanism is exception handling, which can be employed by a programmer to specify how a program should manage an error condition. If a Java program tries to open a file that it has no privilege to read, an exception will be thrown, but the program will not abort. Some of the security-related problems in other languages resulted from programming faults, but the fact that Java is safe cannot protect executing hosts against intentionally malicious programs or smart cards against malicious card applets [5].

10.1.2 Java Type Safety

Java is a strongly typed language. This effectively means that an object must always be accessed in the same way, so that illegal type casting is impossible. By using a cast expression it is possible to instruct a compiler to treat, for example, an integer as a pointer, or a pointer to one type as a pointer to another type. In Java, it cannot happen that one part of the program sees an object as having one type, and another part of the program sees that object as having another type.

Java employs both static and dynamic type checking. Pure dynamic type checking is the safest way to perform type checking. It can be done by checking an object's tag before every operation on it to make sure that the object's class allows such an operation. Unfortunately, dynamic type checking makes programs run slowly. Therefore, Java also employs static type checking, which is much more complicated but can be performed before program execution (i.e., only once). If Java can determine that a particular tag-checking operation will always succeed, then there is no reason to check it dynamically. Static (or load-time) type checking is performed by the bytecode verifier and ensures that the program does not forge pointers, violate access restrictions (i.e., public, protected, private), violate the type of any object, try a forbidden type conversion (illegal casting), or contain stack overflows.

Static checking is performed by the off-card JCVM. Dynamic (or runtime) type checking is performed by the on-card JCVM (i.e., JCRE) and ensures that there are no array boundary overflows or type incompatibilities. Type safety has direct implications on Java security [6].

As pointed out in [7], it would be rather difficult to prove type soundness for Java. Type soundness is based on specifying all possible behaviors that a well-typed program can exhibit, basically by enumerating all errors that may cause the program to abort according to the programming language semantics. In Java many possible reasons exist for run-time errors (e.g., invalid class format), and Java programs may, under some circumstances, terminate in unexpected ways (i.e., cause a segmentation violation).

10.1.3 Transient Objects

Temporary data can be stored in transient objects in RAM. Their contents are set to a default value (e.g., NULL or false) at the end of their lifetime. If the lifetime is defined as CLEAR_ON_RESET, a transient object's contents are set to a default value when the card is reset. If the lifetime is defined as CLEAR_ON_DESELECT, a transient object's contents are set to a default value when the applet is deselected. This feature is very important for security parameters such as the PIN, session keys, or private keys. If such parameters were stored as persistent objects and were not explicitly cleared before the applet was deselected, the applet to be selected next would be able to read their values.

10.1.4 Atomicity of Transactions

An e-payment transaction must be atomic, meaning that it is either fully performed or not at all. In other words, it must not remain in an undefined state. For example, consider the situation in which a card is pulled out of a card reader in the middle of a payment transaction just before the balance on the card is updated but after a valid payment message has been sent to the payee. Without atomicity, this would imply that the payee received the money and would deliver the goods, but the payer's card balance was not reduced. With atomicity, this transaction would simply be aborted.

Java Card supports a transaction model in the following three ways [1]:

1. A single update to a field of a persistent object or a class is always atomic. If an error occurs during update, the content is restored to its previous value.

2. Block updates of multiple data elements in an array are atomic if the arrayCopy method is used.

3. An update of several different fields in different persistent objects performed by an applet can be atomic so that either all updates take place or all fields are restored to their previous values.

10.2   Card Applet Security Mechanisms

There are basically two types of applets [1]:

1. Preissuance applets' classes are burned (or "masked") into ROM at the same time as the JCRE during the manufacturing phase of the card life cycle. They are also called ROM applets. Preissuance applet instances are instantiated in EEPROM by the JCRE. Because they are provided by the card issuers (i.e., by a trusted

source) they may declare native methods. Native methods are written in another programming language and are not subject to Java security checks.

2. Postissuance applets' classes can be downloaded (e.g., into EEPROM) onto the card after the manufacturing phase. For security reasons they are not allowed to declare native methods because their content and behavior cannot be controlled by the JCRE.

An applet can register itself with the JCRE by its AID (applet identifier). Card applet authentication is usually based on the AID, but for improved security it is recommended that a cryptographic mechanism be used in addition. For example, all postissuance applets may need to be signed by the origin so that the digital signature of the cap file can be verified before downloading.

The following sections explain two important card applet security mechanisms: an applet firewall enforced by the JCRE and secure object sharing among applets.

10.2.1 Card Applet Firewall

One of the main advantages of Java Card is that it can host multiple applications, that is, multiple applets can reside on one card. This feature, however, raises security issues of code and data sharing, or in other words, the issues of controlling access to code and data on the card. Applets should not be able to access each other's data. For example, no cardholder would be happy if the tax collecting application on his Java Card could read data from his personal bookkeeping application. Therefore, the Java Card has a mechanism called an *applet firewall*, which means that applets cannot access each other's data unless they explicitly allow it through the Shareable interface. PIN-based cardholder authentication is also supported.

The applet firewall is also a Java Card run-time security check, in addition to Java Card language subset type safety checks (see Section 10.1.2). The "normal" Java programming language allows access to public methods even across packages. The Java Card introduces the concept of a *context*, which represents a separate object space shared by all applets belonging to the same package. Because of the firewall mechanism enforced by the JCRE, an applet may not access objects from a different context. The JCRE has access to all applets and objects created by applets (i.e., all contexts), and all applets have access to global arrays owned by the JCRE, such as the APDU buffer.

Applets gain access to JCRE services through JCRE entry point objects. This means that the public methods of such objects may be invoked from any context. References to temporary JCRE entry point objects (for example, APDU objects or JCRE-owned exception objects) cannot be stored by the invoking applet. References to permanent JCRE entry point objects may be stored and reused. Examples are AID instances created by the JCRE to encapsulate an applet's AID when the applet instance is created [1]. Global arrays are a special type of JCRE entry point object. By using them, applets from different contexts may share only primitive data. The shareable interface mechanism explained in the next section makes object sharing possible among applets from different contexts.

10.2.2 Secure Object Sharing

The first Java Cards based their applet data-sharing policy on access control lists. An access control list defined for each identity which item it could access and with which particular access permissions (e.g., read, write). The items were files, and the identities were defined by means of key files and PINs. That approach did not, however, allow object methods to be shared between different applets, but only data. In other words, it was not possible for an applet to invoke another applet's method [8]. The means of sharing objects between applets was introduced by the Java Card 2.1 specification.

Basically, the Java Card 2.1 object-sharing mechanism also uses access control lists, but this time the identities are established through the unique applet identifiers (AIDs). A card applet with a specific AID may obtain an interface belonging to another applet and thus invoke its methods.

The following explanation of the Java Card object-sharing mechanism is based on [8]. If an applet instance (server applet) wishes to share some methods with applets from different contexts, with the Java Card 2.1 API it does the following:

- The server applet defines a shareable interface PI extending the interface `javacard.framework.Shareable`.

- The server applet defines a class PC implementing the shareable interface.

- The server applet creates an instance PO of class PC.

- The server applet registers with the JCRE by submitting its AID.

Object PO is referred to as the *shareable interface object* (SIO). This mechanism was introduced by the Java Card 2.1 specification. When an applet instance (client applet) wishes to access object PO from the server applet, it performs the following steps:

1. The client applet creates an object reference CO of type PI.

2. The client applet calls a system method `getAppletShare-ableInterfaceObject(ServerAID, byte)` with the AID of the server applet and with an optional byte carrying the identifier of the selected interface (if more than one is provided by the particular server applet). The JCRE forwards the request to the server applet with the first argument replaced by the client applet's AID.

3. When the server applet receives the request, it makes its access control decision based on the client applet's AID; if the client is permitted to share object PO, the server returns a reference to PO (of type SIO), otherwise it returns a `null` to the JCRE.

4. The JCRE forwards the object reference to the client applet; the client casts the object reference to type PI and stores it in CO.

Now when the client applet invokes a method on CO, a context switch is triggered in the JCRE. This means that because of the applet firewall the client can see only the object SO, and the server can see only the arguments passed on the stack (as well as the APDU buffer).

This object sharing model does, however, have some serious security problems:

- *AID spoofing.* Access control decisions made by the server applet are exclusively based on AIDs. If a malicious and fake applet has the AID falsely set to be the same as a client applet known to the server and it (instead of the genuine client applet) is loaded onto the card, it may gain access to the shared interface. The solution to this problem is to allow loading of only applets signed by a trusted source.

- *Inflexible access control.* Because access control is based on AIDs, a server applet must know in advance (i.e., before being loaded onto the card) the AIDs of all applets with which it will share objects. If an applet to share an interface with is written after the server applet has been loaded, there is no flexible way to add the new AID to the server applet's access control list.

- *Illegal reference casting.* Suppose a server applet shares interface PI1 with an applet specified by AID1, and interface PI2 with an applet specified by AID2. Client applet AID1 could, after legitimately obtaining interface PI1, cast interface PI1 into interface PI2 and thus gain access to methods not intended to be shared with it by the server applet. In [8] two work-arounds are proposed: (1) to use a separate delegate object for each shared interface that redirects calls to the intended object, or (2) to check the AID of a client applet each time it tries to access a server applet's method.

- *Inability to pass object parameters.* The only way to pass object parameters between the server and the client is to use the APDU buffer (i.e., global array; see Section 10.2.1). This approach is sometimes very inconvenient because the data to be passed must first be converted into a representation suitable for this type of exchange. Unfortunately, allowing applets to access objects including their data and not only interfaces could potentially open up new security holes.

10.3 Java Card Crypto APIs

Java Card cryptography APIs are based on the Java Cryptography Architecture (JCA),[2] which represents a framework for accessing and developing cryptographic functionality for the Java platform. Because of U.S. export regulations on cryptography it was necessary to provide algorithm extensibility and independence so that different cryptographic algorithms could be implemented by the JCRE providers. In addition, implementation interoperability ensures that applets can access cryptography services on the card without knowing the actual name of the implementation class. This is achieved by factory methods and naming conventions for specifying algorithms and their parameters. For example, SHA can be specified as `Mes-sageDigest.ALG_SHA`; an instance of a class implementing SHA can be obtained by calling the factory method `getInstance(algorithm, externalAccess)` of the `MessageDigest` class. The `algorithm` parameter is set to SHA, and the `externalAccess` parameter can be set to true or false. If it is set to true, the `MessageDigest` instance may be shared among multiple applet instances and it is accessible via a `Shareable`

---

2. http://java.sun.com/products/jdk/1.3/docs/guide/security/CryptoSpec.html.

interface when the owner of the instance is not the currently selected applet (see also Section 10.2).

The two crypto API packages are `javacard.security` and `javacardx.crypto`. The `javacard.security` package contains interfaces for implementing the following:

- Symmetric and asymmetric keys (`Key`, `SecretKey`, `DESKey`, `PrivateKey`, `PublicKey`, `RSAPrivateKey`, `RSAPrivateCrtKey`, `RSAPublicKey`, `DSAKey`, `DSAPrivateKey`, `DSAPublicKey`, `KeyBuilder`, `KeyPair`);
- Authentication (`MessageDigest`, `Signature`);
- Random data generation (`RandomData`);
- Crypto exceptions (`CryptoException`).

The classes in the `javacardx.crypto` package (`Cipher`, `KeyEncryption`) are subject to U.S. export control (strong encryption).

10.4 PIN Verification

To prevent unauthorized use of a smart card, the user is usually required to enter a PIN, an alphanumerical string having six to eight characters at most (otherwise it would be too difficult for the user to remember it). The card owner types his PIN on a PC keyboard or on a keypad on the card reader (i.e., CAD). The keypad is more secure because it is not possible on the PC to intercept the PIN from the keyboard strokes. The card locks after a certain number (e.g., three) of unsuccessful attempts to enter the right PIN. PIN initialization is performed at applet creation and installation [9]. This means that it is possible to define a different PIN for each application (i.e., applet) on the card.

The PIN is represented by a public `PIN` interface in the `javacard.framework` package. An implementation maintains the following values:

- PIN value;
- Maximum number of unsuccessful attempts allowed;
- Maximum PIN length;
- Remaining number of unsuccessful attempts allowed;
- Validated flag (true if a valid PIN has been presented).

References

[1]   Chen, Z., *Java Card Technology for Smart Cards*, Reading, MA: Addison-Wesley, 2000.

[2]   Gosling, J., B. Joy, and G. Steele, *The Java Language Specification*, Reading, MA: Addison-Wesley, 1996.

[3]   Lindholm, T., and F. Yellin, *The Java Virtual Machine Specification*, Reading, MA: Addison-Wesley, 1997.

[4]   MageLang Institute, "Fundamentals of Java Security," Jan. 2000; available at http://developer.java.sun.com/developer/onlineTraining/Security/Fundamentals/index.html.

[5]   Hassler, V., *Security Fundamentals for E-Commerce*, Norwood, MA: Artech House, 2001.

[6]   McGraw, G., and E. Felten, "Java Security and Type Safety," *Byte*, Vol. 22, No. 1, 1997, pp. 63–64.

[7]   Volpano, D., and G. Smith, "Language Issues in Mobile Program Security," *Mobile Agents and Security*, G. Vigna (ed.), LNCS 1419, Berlin: Springer Verlag, 1998, pp. 25–43.

[8]   Montgomery, M., and K. Ksheerabdhi, "Secure Object Sharing in Java Card," *Proc. USENIX Workshop on Smartcard Technology*, Chicago, IL, May 10–11, 1999; available at http://www.usenix.org/publicaitons/library/proceedings/smartcard99/montgomery.html.

[9]   Chen, Z., "How to Write a Java Card Applet: A Developer's Guide," *JavaWorld*, July 1999; available at http://www.javaworld.com/jw-07-1999/jw-07-javacard.html.

11

Application Development

This chapter gives an introduction to Java Card application development, beginning with an overview of core classes and methods of the Java Card API. The API is explained on a general level. Readers interested in details of particular classes or methods should refer to the Java Card API reference manuals. The section concludes by presenting Java Card implementations currently available on the market.

11.1 Java Card API

The Java Card API , also referred to as the *framework*, consists of four core packages. Two packages, called `java.lang` and `javacard.framework`, were mentioned in previous sections of the book. The `java.lang` package contains all classes related to support of the Java programming language subset. The `javacard.framework` package contains classes related to Java Card applet functionality.

The other two packages, `javacard.security` and `javacardx.crypto`, play a particularly important role for several reasons. First of all, they contain classes related to the security functionality of an applet. This security functionality mainly covers the cryptographic API and its support. Unfortunately, implementation of the security-related Java Card classes is still not standardized among smart card manufacturers. Usually, smart card manufactures provide security-related classes within their own proprietary packages, which are shipped in a form of an extension API. Therefore, their description is omitted in the sections that follow (see also Section 10.3). However, a closer look at manufacturer-specific issues of the Java Card API will be taken in Section 11.2, where existing Java Card implementations are presented.

The following sections provide an overview of the core classes of the `javacard.framework` package.

11.1.1 JCSystem Class

The `JCSystem` class contains a number of methods for applet execution control, object management, and atomic transaction support. All methods of the class are static. A rather large group of the class methods is used for performing atomic transactions (see Section 10.1.4). The methods `begin-Transaction()`, `abortTransaction()`, and `commitTransaction()` are dedicated to starting, aborting, and committing an atomic transaction, respectively. Some other methods provide additional data on atomic transactions, for example, the transaction depth, the memory used, and the memory still available in the transaction commit buffer.

Methods of the group `MakeTransient...Array` are used to create transient arrays of the types `boolean`, `byte`, `short`, and any other custom object type. Apart from specifying the number of elements in an array, these methods identify an event determining when the array elements are cleared (see Section 10.1.3). The method `isTransient` may be used to verify whether the given object is transient or not, and, if yes, to determine the type of event on which the content of the object is cleared to a default value.

Other class methods are rather specific, and their detailed description can be found in the Java Card API reference manuals.

11.1.2 Applet Class

The abstract class `Applet` must be implemented by any Java Card applet. The class inherits all necessary functionality for an applet's installation, registration, and execution. The core methods of the class were discussed in Section 8.3; Table 11.1 gives a brief summary of these core methods.

Two other class methods should also be mentioned. The purpose of the method `selectingApplet()` is quite interesting. It returns a boolean value indicating whether an applet has just been selected or not. The method is used in an applet `process` method in order to distinguish the applet SELECT APDU from any other SELECT APDUs that might be sent to the applet (see Figure 8.3).

Table 11.1
Summary of the Class `Applet` Core Methods

Method	Description
install	Called by JCRE in order to create an instance of an applet.
register	Invoked by an applet in order to register itself within JCRE.
select()	Called by JCRE to inform an applet that it was selected. Must return true to indicate a successful selection.
deselect()	Called by JCRE to inform an applet that either another applet was selected or the applet was selected again.
process	Called by JCRE in order to process an incoming APDU. The APDU itself is given as a method parameter.

The `getShareableInterfaceObject` method is called by JCRE in order to obtain a shareable interface object from this applet. For details related to object sharing between applets run on one card, see Section 10.2.2.

11.1.3 APDU Class

The APDU class encapsulates a complete set of features needed to process an incoming APDU, to prepare an outgoing response APDU, and to send it out. The APDU object is owned by JCRE. An applet receives its instance as a parameter of the applet `process` method. The APDU object has a byte array buffer that is used for storing both header and data bytes of incoming APDUs and data bytes of response APDUs.

We should mention that the class is dedicated only to command and response APDUs built in accordance with ISO/IEC 7816-4. Another convenient feature of the APDU class is that its methods apply to any (T = 0 or T = 1) communication protocol used by a smart card.

When the APDU object is passed by JCRE to the `process` method of an applet, its buffer array already contains the incoming APDU header, that is, the CLA, INS, P1, P2, and L$_c$ bytes. The reference to the APDU byte array buffer can be obtained with the class method `getBuffer()`. To place data bytes, if any, in the buffer array, another method of the APDU class is called: the `setIncomingAndReceive()` method. A remarkable property of this method is that it places only as many bytes to the array buffer as can securely fit there, thereby avoiding buffer overflow.

If the incoming APDU contains more data bytes than will fit the array buffer (as it was copied by the setIncomingAndReceive() method), the remaining bytes can be placed to the buffer using subsequent calls of the receiveBytes method. This method also copies to the buffer only as many bytes as it will fit there.

The following example demonstrates the basics of reading APDU header and data bytes:

```
public void process(APDU apdu) {
    // get the reference to the array buffer
    byte[] buffer = apdu.getBuffer();
    // read CLA and INS bytes from the buffer
    byte cla  =  buffer[ISO7816.OFFSET_CLA];
    byte ins  =  buffer[ISO7816.OFFSET_INS];
    ...
    // incoming APDU has some data bytes; read them
    short bytesRead = apdu.setIncomingAndReceive();
    ...
}  // process(APDU apdu)
```

Abstract interface ISO7816 contains static constants defining offsets of particular APDU bytes in the array buffer. Naturally, CLA, INS, P1, P2, and L_c have offsets 0, 1, 2, 3, and 4, respectively. The offset of the beginning of the data bytes is 5 and is defined by the static constant OFFSET_CDATA of the interface ISO7816.

Generating a response APDU is as easy as processing a command APDU. The class has a number of methods for preparing and sending out a response APDU. The easiest and most convenient method is setOutgoingAndSend. This method prepares a response APDU and sends it out immediately. It is used when all response APDU data fit into the array buffer. For instance, sending two bytes of response can be accomplished in the following way:

```
buffer[0] = (byte) 0xA0;
buffer[1] = (byte) 0xB0;
setOutgoingAndSend( (short) 0, (short) 2);
```

The first parameter of the method specifies an offset of the outgoing data in the array buffer (note that the same buffer is used for receiving and sending). The second parameter specifies the length of the outgoing data. As a result of the method execution, JCRE will send out a response APDU with two data bytes and the success status word 90 00 H automatically attached.

The same can be accomplished by using three other methods sequentially: `setOutgoing()`, `setOutgoingLength`, and `sendBytes`. To send large response APDUs, the method `sendBytesLong` is used.

11.1.4 OwnerPIN Class

The `OwnerPIN` class is an implementation of the `PIN` interface comprising functionality related to PIN verification (see Section 10.4). The class maintains the PIN value, a maximum number of unsuccessful tries allowed, and a try counter. If the maximum number of unsuccessful tries exceeds the defined limit, the `PIN` is *blocked*, meaning that even at this point presenting a correct PIN value will not result in successful PIN verification.

The class maintains the PIN validation flag indicating whether the PIN was successfully verified since the last reset of the card or of the `OwnerPIN` object. The validation flag is stored in the volatile memory of a card, which guarantees that its value is cleared after each card reset. The value of the validation flag is accessible via the method called `isValidated`.

Another class method `check` allows the PIN value to be verified against a given value. If the correct value is presented, the try counter is reset to its maximum value and the validation flag is set to true. The method returns a boolean value indicating whether the PIN verification was successful or not.

Other class methods make it possible to unblock the PIN (`resetAndUnblock`), to change the PIN value (`update`), to retrieve the remaining number of tries until the PIN will be blocked (`getTriesRemaining`), and to reset the PIN (`reset`).

11.1.5 Util Class

The `Util` class provides a number of handy utility functions that may be needed by an applet. Java Card specifications allow some of those functions to be implemented as native methods in order to increase their performance. As all methods of the class are static, the class does not need to be instantiated. Instead, the JCRE instance of the class can be used to invoke a desired method.

Basically, the class consists of two groups of methods. The first group contains several methods for operations on arrays: comparing two byte arrays (`arrayCompare`), copying byte arrays atomically (`arrayCopy`) and non-atomically (`arrayCopyNonAtomic`), and filling a byte array with a given byte value nonatomically (`arrayFillNonAtomic`).

The second group contains methods that deal with conversions between types `short` and `byte`. The methods make it possible to construct a `short` value from two `byte` values given separately or in a byte array, and to divide a `short` value into two `byte` values.

11.1.6 Interface `ISO7816`

Interface `ISO7816` contains a wide range of constants related to protocols and data structures defined in ISO/IEC 7816-3 and ISO/IEC 7816-4. In general, the constants fall into the following main groups:

- Constants defining an offset of particular bytes (e.g., CLA, INS, etc.) in the APDU buffer;
- Constants defining CLA and INS codes in accordance with ISO/IEC 7816-4;
- Response status word codes defined in accordance with ISO/IEC 7816-4.

Each of the constants can be accessed through the respective JCRE instance.

11.2 Existing Implementations

During the last few years, Java Card has met with good support from smart card manufacturers, and a number of Java Card implementations have appeared on the market. Apart from the standard Java Card API, each manufacturer also provides some extension API. This section presents three well-known Java Card implementations and discusses their special features. Table 11.2 at the end of the chapter presents an overview of the main characteristics of the Java Card implementations discussed in this section.

11.2.1 Giesecke & Devrient Sm@rtCafé

The Sm@rtCafé 1.1 card from the German company Giesecke & Devrient,[1] though based on the Java Card 2.1 specifications, is not fully compliant with Java Card 2.1. The card supports both T = 0 and T = 1 protocols that are selectable through protocol type selection (PTS). It is implemented on a single-chip microcontroller of the Siemens SLE66 family. The card has

---

1. http://www.gdm.de.

1,280 bytes of RAM, 32 Kbytes of ROM, and from 8 to 16 Kbytes of EEPROM for operating systems, applications, and data. The card chip is based on an 8-bit architecture and operates on the 7.5-MHz frequency.

The Sm@rtCafé 1.1 card exists in two variations: standard and crypt. The crypt version extends the basic Java Card functionality of the standard version by a comprehensive cryptographic API delivered as separate packages: `com.gieseckedevrient.javacardx.crypto` and `com.gieseckedevrient.javacardx.cryptox`.

Basically, the cryptographic API provides support of the following algorithms and services:

- DES and DES3 algorithms;
- RSA algorithm with the key length up to 1,024 bits;
- Secure hash algorithm SHA-1;
- External and mutual authentication services based on the DES algorithm;
- Digital signature service according to ISO/IEC 14888-3;
- Session key derivation service.

The functionality of each service and algorithm is encapsulated in a separate class. For instance, the `Authentication` class provides all necessary functionality for performing external or mutual card terminal authentication. The class methods make it possible to request challenge data from a card, to generate a session key, and to perform mutual or external authentication by simple methods invocation. In a similar manner, classes `Signer` and `Verifier` include the functionality needed to generate and verify, respectively, RSA digital signatures.

In addition, Sm@rtCafé 1.1 cryptographic API contains the `SecureRandom` class, which provides a source of *cryptographically* secure random numbers; that is, random numbers that cannot easily be guessed, predicted, and so on. This feature is very important for security of challenge-response authentication services or session key derivation services.

The installation program of the Sm@rtCafé 1.1 card, called the Main Loader, allows the definition of four security levels for loading an applet:

1. An applet is loaded in plaintext.
2. A loaded applet is digitally signed and the signature is verified by the Main Loader.

3. A loaded applet is encrypted.

4. A loaded applet is both digitally signed and encrypted.

The Main Loader can be configured in such a way that further reconfigurations are not allowed and the card application memory area cannot be cleared. This ensures that a loaded applet (or applets) cannot be deleted from the card or replaced.

The development environment that comes with the Sm@rtCafé 1.1 development toolkit makes it possible to perform complete simulation of the Sm@rtCafé 1.1 Java Card virtual machine and to perform applet bytecode-level debugging and tracing.

11.2.2 Gemplus GemXpresso 211

GemXpresso 211 is a family of smart card products from the Gemplus company.[2] GemXpresso 211 V2 is the currently available representative of this family and is the first Java Card technology smart card that complies with both Java Card 2.1 specifications and Visa Open Platform 2.0 specifications. The card supports both $T = 0$ and $T = 1$ protocols; the $T = 1$ protocol is available only after a warm reset of the card. The card is based on the 8-bit microcontroller and has 32 Kbytes of ROM, 32 Kbytes of EEPROM, and 2 Kbytes of RAM.

The security of GemXpresso 211 V2 and its components has been evaluated and certified by a number of international bodies. The card itself was certified by Visa with the highest security level 3. Security of the previous version of the card was evaluated according to the Common Criteria and received the assurance level EAL1. Security of the card cryptographic hardware components was evaluated according to the Common Criteria and received the assurance level EAL3.

Cryptographic support of the GemXpresso 211 V2 is limited only to DES and DES3 implementation. The RSA algorithm, secure hash algorithms, and digital signature algorithms are not supported. Also, the card provides no support for authentication services. In accordance with Java Card specifications, cryptographic functions are placed in the packages javacard.security and javacardx.crypto.

Implementation of the Visa Open Platform specification makes it possible to establish a secure communication channel between a GemXpresso

---

2. http://www.gemplus.com.

211 V2 card and a card terminal on the APDU level. A secure channel ensures APDU integrity and confidentiality as well as communication session authenticity. Visa Open Platform API is provided in the form of a separate package `visa.openplatform`.

We should mention that the next-generation card of the GemXpresso 211 family, GemXpresso 211/PK, will support the RSA algorithm and secure hash algorithms MD5 and SHA.

The GemXpresso 211 V2 toolkit is supplied with the GemXpresso RAD 211 development environment. It allows simulation of applet execution on the card and performance of debugging.

11.2.3 Schlumberger Cyberflex Access

The Schlumberger company was the first smart card manufacturer to release a smart card programmed in the Java programming language. The latest Java Card technology smart card from Schlumberger is Cyberflex Access.[3] Cyberflex Access is compliant with an earlier version of Java Card, namely, Java Card 2.0. The card supports only T = 0 protocol and has 16 Kbytes of EEPROM.

The cryptographic facilities of Cyberflex Access include the following:

- DES and DES3 algorithm implementation;

- RSA algorithm implementation with the key size up to 1,024 bits;

- External and internal card terminal authentication services;

- SHA-1 secure hash algorithm implementation.

The cryptographic API that covers the implementation of the algorithms mentioned above is located in the `javacardx.crypto` package.

An interesting feature of Cyberflex Access is that, in contrast to other Java Card technology smart cards, it supports the ISO/IEC 7816 file system. The files are accessed and managed via the Loader application of a Cyberflex Access card. The Loader application can be regarded as an extended variation of the Java Card installation program. Besides loading applets, the Loader also supports commands for file access and management and basic security mechanisms, such as card holder verification (CHV).

---

3. http://www.cyberflex.slb.com.

Cyberflex Access is supplied with an extension API delivered as a separate package, javacardx.framework. The extension API provides classes and methods for card-specific APDU processing (CyberflexAPDU class), operations related to files (CyberflexFile class), and native operating system calls (CyberflexOS class).

Table 11.2 summarizes the Java Card implementations just discussed.

Table 11.2
Summary of Java Card Implementations

	Sm@rtCafé	GemXpresso 211	Cyberflex Access
Manufacturer	Giesecke & Devrient	Gemplus	Schlumberger
Resources	1,280 bytes RAM, 32 Kbytes ROM, up to 16 Kbytes EEPROM	2 Kbytes RAM, 32 Kbytes ROM, 32 Kbytes EEPROM	16 Kbytes EEPROM
Supported protocols	T = 0, T = 1	T = 0, T = 1	T = 0
Java Card version	Java Card 2.1	Java Card 2.1	Java Card 2.0
Other specifications	—	Visa Open Platform 2.0	—
Cryptographic algorithms	DES, DES3, RSA SHA-1	DES, DES3	DES, DES3, RSA SHA-1
Security services	External and mutual authentication, ISO/IEC 14888-3 digital signature, session key derivation	—	External and internal authentication

Part III
OpenCard Framework

The OpenCard Framework (OCF) is one of the most important platform-independent smart card application interfaces with real multiapplication support. Chapter 12 describes the OCF, its structure, and the related technologies with an emphasis on the PC/SC. Chapter 13 continues with a discussion of basic OCF functionality and the OCF security concept. In addition, it briefly describes how the OCF communicates with a card terminal and how it can be used with Java Card applets.

12

OCF Basics

This chapter explains the basic background concepts that have led to the development of the OCF. The main goal of the framework is to provide an easy-to-use high-level API for smart card–aware applications running on the terminal. The OCF is an object-oriented framework that uses some well-known design patterns. Not surprisingly, the OCF architecture is built around the already established distribution of functionality among the main players in the smart card world, namely, the terminal application provider, card operating system provider, card issuer, and card terminal provider. Section 12.1 explains the needs of smart card–aware terminal applications. These needs can be fulfilled by the OCF, as pointed out in Section 12.2. Section 12.3 gives a brief overview of the OpenCard Consortium development. In Section 12.4 the OCF architecture is depicted, including explanations of some important design patterns it employs (framework in Section 12.4.1 and abstract factory in Section 12.4.2). Chapter 13 gives more details about the OCF.

Another important interindustry initiative aimed at card terminal interoperability is the personal computer/smart card (PC/SC), which is presented in Section 12.5. Section 12.6 compares the two approaches (OCF and PC/SC) from different perspectives. They have many similarities, but also some differences of which the designer must be aware before deciding which concept to use. Finally, to make the picture of the card terminal

123

interoperability initiatives complete, Section 12.7 gives a brief overview of several card terminal APIs.

12.1   Smart Card Applications

Smart card applications generally consist of two parts, one part running on the card and one part running on the application terminal. Both parts may be developed by a single application provider, but the card issuer decides where on the card to place the card-resident part. In the early stages of smart card application programming, the terminal application developer was supposed to know the card specifics (such as the card operating system and the communication protocol supported by the card) in order to use the card in his application. These issues are now addressed by international standards so that any application conforming to the ISO 7816 protocols can interoperate with a standard smart card (see Section 1.5). Unfortunately, the exact definitions of the set of command APDUs and response APDUs may differ.

As of 2001, there is no standardized API for communication with cards. In other words, if a card terminal developer wishes to support different cards with the same functionality but with different proprietary APIs, he must either address each API in his application, or study the ISO standards describing communication protocols and card operating system commands, and possibly even the card manufacturer's documentation describing proprietary commands, and incorporate them in his design.

For communication with a card, a card acceptance device (called a *card terminal* in OCF parlance, e.g., a card reader) is needed. For a card terminal to be used, the corresponding card terminal driver must be installed on the application terminal. Some card terminals just pass the data to and from the card, and some may implement a more complex functionality to interpret and act on the data sent from the application terminal to the card, and even perform some additional operations. In addition, a card terminal may be equipped with a display, a PIN pad (e.g., Sign@tor by Siemens AG Austria, shown in Figure 12.1), or even a biometric input device. Card terminals from different vendors may thus exhibit different functionality and support different communication protocols. Moreover, card terminals can be attached to different I/O ports (e.g., serial port, PC Card bus). The fact that communication with card terminals is not standardized leads to portability problems, because a terminal application developer must specifically consider each card terminal and its control commands and determine which should be supported by his application.

Figure 12.1 A Siemens Sign@tor Version 1.0 card terminal with a display and a PIN pad. (*Source:* Siemens AG Austria.)

12.2 The OCF

The OCF addresses the problems mentioned in the previous section. It represents a set of open technical specifications defining a Java-based framework for developing platform-independent smart card applications. The main objectives are as follows [1]:

- To "hide" the parts specific to the card terminal vendor, the card operating system, and the card issuer;
- To allow faster and cheaper terminal application development by providing a high-level API;
- To allow portability of terminal applications to many different platforms.

These objectives ensure that differences or changes in the card operating system, card terminal, or application management scheme do not affect the terminal application. The platforms that have been used for OCF development, testing, and demonstrations so far are WinNT, Win95, Linux, IBM AIX, and several network computers.[1]

---

1. http://www.opencard.org/misc/OCF-FAQ.shtml.

12.3   The OpenCard Consortium

The OCF specifications are issued by the OpenCard Consortium, a non-profit, U.S.-based organization whose members are the leading companies involved in the smart card business (IBM, Gemplus, Gisecke & Devrient, Schlumberger, Sun Microsystems, Visa, and others). The interindustry initiative began in 1997 when a small group of companies started to develop a Java framework for smart card deployment. Those efforts resulted in the first version of the OCF in 1998. Shortly thereafter the OpenCard Consortium was founded to promote the OCF, support its further development, and allow compliance testing.

The OpenCard Technical Committee takes care of the development of OCF specifications. The work process is based on two types of documents:

1. Requests for proposals (RFPs) propose a new work item to the committee, which can be a change, an extension, or an enhancement to the specifications.

2. Requests for comments (RFCs) propose a solution to a problem stated in a particular RFP.

An RFC must be approved by the technical committee and the management board in order to be officially integrated into the OCF. RFPs and RFCs that have been voted on are public and can be obtained from the OCF Web page [2].[2]

To make smart card transactions possible on "tiny devices" with limited resources, such as payment terminals, payphones, or set-top boxes, it may be necessary to reduce an API's size at the expense of its flexibility. Consequently, for such platforms the framework requirements may be rather different. For example, some OCF components and mechanisms could be omitted if not of crucial importance for the functionality. The Tiny OCF (TOCF) is a version specially designed for embedded devices [3].

12.4   OCF Architecture Overview

Figure 12.2 shows the OCF architecture. As illustrated by their color (white, light gray, medium gray, and dark gray), the boxes come from different sources:

---

2. http://www.opencard.org.

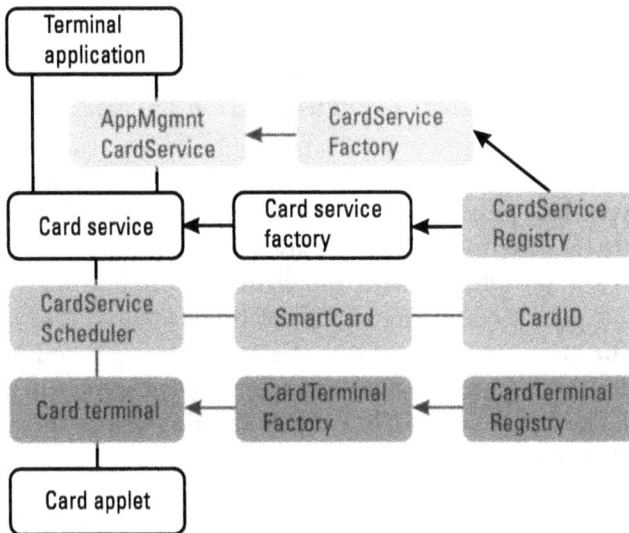

Figure 12.2 OCF architecture overview.

- The white boxes represent the on-card and the off-card parts of the application, including the API. The Terminal Application box is provided by the application developer. The CardService box and the CardServiceFactory box as well as the Card Applet box are provided by the on-card application (i.e., card applet) developer. In Part IV we demonstrate how to develop the white boxes.

- The light gray boxes are provided by the card issuer.

- The medium gray boxes are provided by the OCF (i.e., they are contained in the OCF core).

- The dark gray boxes are provided by the card terminal vendor.

What follows is a brief description of all boxes shown in the figure [4]. As previously mentioned, the Card Applet is provided by the card developer and runs on a smart card.

The CardTerminal classes are provided by manufacturers who wish to make their card terminals available to the OCF-compliant terminal applications. A CardTerminal class encapsulates the card terminal behavior. The CardTerminalFactory is used by the OCF to create CardTerminal instances when the framework is initialized. The card terminal factory of each card terminal attached to the desktop computer has to be registered with the `CardTerminalRegistry`. To allow modeling of card terminals with more than one slot for card insertion, a `CardTerminal` can have one or more objects of type `Slot`. A card terminal can also have a PIN pad and a display.

The `SmartCard` class is assigned a `CardID` that uniquely identifies the card type. This piece of information is obtained when the smart card is inserted into the card reader (ATR; see also Section 5.1).

The OCF provides methods and classes for `CardService` to access the card (e.g., `CardChannel`, `CardServiceScheduler`). Because there may be more than one instance of a card service per card, the `CardServiceScheduler` serializes the access of different services to the `CardChannel`. The `CardChannel` is a communication link to the card that is represented by the `CardID`. Consequently, there is one `CardServiceScheduler` object for each card.

The `CardService` classes implement a standard API, thus hiding the smart card specifics. It generates APDUs and communicates with the card to support high-level API functions. A `CardServiceFactory` is associated with each `CardService` implementation and is capable of constructing it. The `CardServiceFactory` identifies the card or cards for which the `CardService` was designed. When a smart card is inserted into the reader, the OCF goes through its list of registered card service factories (within the `CardServiceRegistry`) and instantiates card services corresponding to the card. When a `CardService` with a particular interface is requested, the `CardServiceRegistry` calls every `CardServiceFactory` registered for a specific `CardID` of the `SmartCard` object until an appropriate `CardService` has been created. Card services can use functions from other card services.

Currently, the OCF defines only a few standard card service interfaces (e.g., `FileAccessCardService` or `AppletManagerCardService`; see Section 13.2.3) to make the common smart card functions available to the application programmer. Smart card issuers are supposed to provide implementations of these classes and the corresponding factory classes.

The Terminal Application is written by the application programmer, who only needs to know the API provided by the `CardService`. More details about the structure of the framework classes will be given in Chapter 13.

As can be seen from the previous description, the central OCF design patterns are the framework and the factory, which are explained in the following two sections.

12.4.1 What Is a Framework?

As pointed out in [2], the problems described in the previous section can be solved by an object-oriented *framework* approach. A framework is more than a simple class library. With a library, the applications invoke operations implemented by the library or instantiate classes implemented by the library. With an object-oriented framework, the applications can also use the operations, classes, and interfaces provided by the framework, and can extend the framework by following its principles and coding and naming conventions. For example, an approach to supporting card terminals from different manufacturers is to provide an interface or an abstract class for card terminals that can be subclassed to implement a driver for a particular card terminal.

In addition, a framework can call operations implemented by the application and even control the flow of execution. In other words, reuse leads to an inversion of control between the application and the framework [5]. When using a conventional library, the application developer writes the main body of the application and calls the code she wants to reuse. When using a framework, the developer reuses the main body (i.e., the core of the framework) and writes the code the framework calls. In this way the developer is freed from having to make design decisions and is supposed to write methods according to the framework-specific names and calling conventions.

Obviously, with a framework, application development becomes much faster and the applications have similar structures, which in turn leads to easier maintainability and better readability. On the other hand, the developer loses some creative freedom, because many design decisions have already been made by the framework. In addition, because applications are so dependent on the framework, they are very sensitive to any changes in the framework interface. This means that applications are actually forced to evolve with the framework [5].

Obviously, it is very difficult to design a successful framework. Here are some requirements that a good framework should satisfy [5, 6]:

- Completeness, which means that it supports features needed by the applications. For example, a smart card framework should implement the functionality defined by the ISO 7816 standards (communication protocols, basic operating system commands, application selection), and by the Open Platform operating systems (Java Card, MULTOS, WfSC) [2].

- Reusability, meaning that its interfaces and abstract classes can be used in different contexts. A framework based on well-known design patterns is more likely to achieve high levels of design and code reuse.

- Extensibility, so that it is easy to modify the existing functionality or add new functionality by deriving new classes.

- Understandability, which means that it is well documented and follows the standard design and coding guidelines.

- Flexibility, so that the changes in the framework imply as few changes in existing applications as possible.

12.4.2 What Is an Abstract Factory?

As explained in the previous section, frameworks use abstract classes to define and maintain relationships between objects. Abstract factories are a design pattern that provide an interface for creating families of related objects without specifying their concrete classes [5]. This design pattern is very suitable for the OCF because the terminal applications are supposed to support different types of smart cards, card terminals, and card services (i.e., on-card applications). In other words, terminal applications should not "hard code" the cards, the card terminals, or the card services.

In general, the advantages of abstract factories are as follows [5]:

- A factory encapsulates the responsibility and the process of creating product objects and in this way isolates applications from implementation classes. Product class names are isolated in the implementation of the concrete factory, so they do not appear in application code.

- Because the class of a concrete factory appears only once in an application, it is easy to change the concrete factory that an application uses. The application can use different product configurations simply by changing the concrete factory.

- When product objects in a family are designed to work together, using an abstract factory can enforce the use of objects from one family at a time.

Let us consider an example. In the OCF an on-card application developer is supposed to supply a class for his specific card service (see Section 12.4).

When a card service with a particular interface is requested from the terminal application, the OCF registry of card services needs a way to instantiate the proper card service depending on the smart card type. A "card service factory" is associated with each card service implementation and is capable of constructing it. The OCF uses abstract factories in the following way:

- `CardServiceFactory` represents an abstract factory. It declares an interface for operations that create card service objects.

- `SpecificCardServiceFactory` represents a concrete factory implementing the operations to create specific card service objects.

- `CardService` is an example of an abstract product. It declares an interface for a type of card service object.

- `SpecificCardService` defines a card service object to be created by the corresponding specific card service factory and implements the card service interface.

- Finally, the terminal application uses only interfaces declared by `CardServiceFactory` and `CardService` classes.

A similar mechanism is implemented for card terminals as well:

- `CardTerminalFactory` declares an interface for operations that create card terminal objects.

- `SpecificCardTerminalFactory` represents a concrete factory implementing the operations to create specific card terminal objects.

- `CardTerminal` declares an interface for a type of card terminal object.

- `SpecificCardTerminal` defines a card terminal object to be created by the corresponding specific card terminal factory and implements the card terminal interface.

The principle of the abstract factory is shown in Figure 12.3 as a simplified Universal Modeling Language (UML) class diagram. The arrows represent inheritance. The dashed arrow represents multiple inheritance because there may be multiple concrete products from a single family (but only one factory that creates them).

An abstract factory (e.g., `CardTerminalFactory`) only declares an abstract interface for creating products (e.g., card terminals). It is the

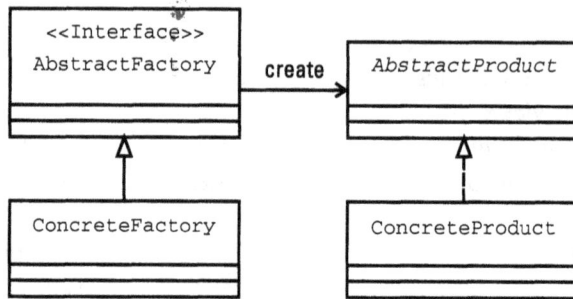

Figure 12.3 The principle of the abstract factory in the OCF (UML class diagram).

concrete factory subclasses (e.g., `SpecificCardTerminalFactory`) that actually create them. The way to implement this in the OCF is to define a factory method for each product (e.g., card service or card terminal). A concrete factory specifies its products by overriding the factory methods for each of them. In other words, a card terminal vendor must provide his own version of a `CardTerminalFactory` that can produce appropriate `CardTerminal` objects.

The main disadvantage of this approach is that supporting new kinds of products requires extending the factory interface, which involves changing the abstract factory class and its subclasses [5].

12.4.3 Singleton and Registry

The purpose of a *singleton* is to ensure that a class has only one instance and to provide a global point of access to it. This works in such a way that the singleton itself is responsible for keeping track of its sole instance. It can ensure that no other instance can be created by intercepting requests to create new objects and provide a way to access the instance [5].

In the OCF, a singleton is used as follows. The `CardTerminalRegistry` is a singleton object that keeps track of all card terminals known to the OCF. It provides methods that allow the addition and removal of devices from the system. The purpose of a registry, which is also a well-known design pattern, is to select a factory at run time. This makes it possible to register instances of new factory subclasses without modifying the existing code. For example, the `CardTerminalRegistry` provides a mapping between string names of card terminals and the `CardTerminalFactory` objects, so it creates factory objects as needed. Each

`CardTerminalFactory` in turn keeps track of the `CardTerminal` classes it instantiated.

Each registry must know at least one factory. When the OCF starts up, the card terminal registry and the card service registry are initialized on the basis of the user-defined properties (from the `opencard.properties` file).

12.5 PC/SC

The PC/SC is another framework aimed at smart card interoperability based on a set of specifications issued by the PC/SC Workgroup.[3] The workgroup, which was founded in 1996, is a collaborative effort of a number of international personal computer and smart card companies. A company may join as a core member (e.g., Microsoft, Hewlett-Packard, Toshiba, Intel) or as an associate member (e.g., Philips Semiconductors, Siemens, iD2 Technologies, ActivCard). The core members serve as a board of directors, whereas associate members provide input and expertise. PC/SC builds on existing industry smart card standards, such as ISO 7816 (compatible with EMV and GSM) and complements them by defining low-level device interfaces and high-level C or C++ APIs, as well as resource management to allow multiple applications to share smart cards attached to a system.

By using the smart card APIs and device management infrastructure, terminal application developers can develop smart card applications that will work with any PC/SC-compliant smart card reader. For smart card manufacturers, PC/SC standards simplify development of interface libraries. Finally, card acceptance device (called an interface device or IFD in PC/SC terminology) vendors benefit because any PC/SC-compliant smart card application will work with a PC/SC-compliant reader.

The first version of the PC/SC specifications was issued in December 1997 [7]. Although the specifications are, in principle, platform independent, some implementations are available only for PCs running a 32-bit Microsoft Windows operating system. There is also an implementation of the Resource Manager for Linux [8].[4] Version 2.0 was announced in November 1999 [9] but was not yet publicly available as of February 2001. Version 2.0 will add the application selection mechanism for multiple-application cards, support for IFDs with extended capabilities (e.g., PIN pad, display

---

3. http:// www.pcscworkgroup.com.
4. http://www.linuxnet.com/middle.html.

and multislot), support for contactless cards, and support for synchronous protocol cards.

The general PC/SC architecture is shown in Figure 12.4. The smart card is referred to as the ICC. PC/SC Version 1 deals with contact-type ICCs as defined by ISO/IEC 7816. ISO 7816-10–compliant ICCs (synchronous cards) are supported as well.

As mentioned before, the card acceptance device is referred to as the ICC IFD. An I/O channel for binary data is established by the electrical connections between the embedded microprocessor of the ICC and the IFD. In addition, through the electrical connections, the IFD provides the microprocessor chip with DC power and a clock signal, which is used to drive the program counter of the microprocessor. An IFD may use a keyboard port, a serial line port, or a PC Card (PCMCIA) port. In the future, the USB will also be supported.

The IFD handler is the layer primarily responsible for facilitating the interoperability between different IFDs. It encompasses the PC software that supports the specific I/O channel between the IFD and the PC and provides

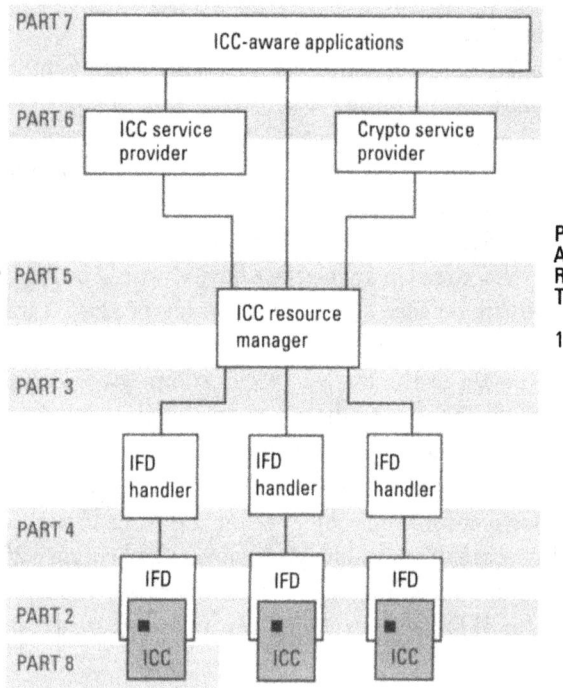

Figure 12.4 PC/SC general architecture.

access to specific functionality of the IFD for the terminal application. The IFD handler supports the ISO 7816-3 communication protocols (T = 0, T = 1) and the synchronous protocol specified by ISO 7816-10.

The ICC resource manager is a system-level component of the architecture that will most likely be provided by the operating system vendor. It runs as a trusted service in a single process. All requests for smart card access go through the resource manager and are routed to the IFD containing the requested ICC. In other words, the resource manager is responsible for managing and controlling all application access to any ICC inserted into any IFD attached to a PC. The resource manager performs three basic tasks in managing access to multiple readers and cards:

1. It identifies and tracks resources (i.e., ICCs and IFDs).
2. It controls the allocation of IFDs and resources (including access to ICCs) across multiple applications.
3. It supports transaction primitives for accessing services available on a specific ICC. This is important because current ICCs are single-threaded devices that often require execution of multiple commands to complete a single function. Transaction control allows multiple commands to be executed without interruption, ensuring that intermediate state information is not corrupted.

The service provider is responsible for encapsulating functionality offered by a specific ICC (e.g., file access, authentication, or cryptographic services) and making it accessible through high-level programming interfaces. It is divided into two independent components: the ICC service provider and the cryptographic service provider. Only the ICCs with cryptographic functionality that should be made accessible to application programs need a cryptographic service provider. The PC/SC specification defines interfaces for general-purpose cryptographic services including key generation, key management, digital signatures, hashing, bulk encryption, and key import/export.

A service provider may also expose interfaces that use cryptography internal to the ICC, such as secure messaging or cryptogram-based authentication. An application need not know in advance which service provider it wants to use, but may determine it at run time by using the ICC resource manager to enumerate the available providers and their supported interfaces. Before a service provider can be used, it must be registered with the ICC resource manager, usually through an ICC setup utility provided by the ICC vendor.

Finally, the ICC-aware application is the terminal-resident ICC application. It is assumed that the application is running as a process within a multiuser, multiprocess, multithreaded, and multiple-device environment. The components defined within the PC/SC specification for Version 1.0 provide mechanisms to map the application requests to the ICC, which is typically a single-user, single-threaded, but multiapplication environment.

Figure 12.4 also shows how the different parts of the PC/SC specification are related to the overall system architecture:

- Part 1 provides an overview of the system architecture and components.

- Part 2 describes compliant ICC-IFD characteristics and interoperability requirements.

- Part 3 describes the interface to, and required functionality for, compliant IFDs.

- Part 4 discusses design considerations for IFDs and provides a recommended implementation for IFDs with an integrated PS/2 keyboard.

- Part 5 specifies the interfaces and functionality supported by the ICC resource manager.

- Part 6 describes the ICC service provider model, identifies required interfaces, and indicates how this model may be extended to meet domain-specific requirements.

- Part 7 describes design considerations for application developers and the usage of other components.

- Part 8 describes recommended functionality for ICCs intended to support general cryptographic and storage requirements.

12.6 OCF Versus PC/SC

The PC/SC interface and the OCF are both interindustry initiatives aimed at defining a standard way to integrate smart cards into computer systems. Interoperability for smart cards means that one vendor's card terminal (e.g., card reader) or card can be replaced with another manufacturer's card terminal or card. With respect to their scope and to the target deployment environments, these efforts are complementary rather than competing. Because they both address the communication between card terminals and smart cards, there is some overlapping between them. Also, they have some concepts and mechanisms in common. The following sections give a brief comparison between the OCF and the PC/SC [10, 11].

12.6.1 Platform

In the PC/SC much emphasis was placed on the interoperability of smart cards and card terminals, and on the integration of those card terminals into the Microsoft Windows operating system. The founding members of the OpenCard Consortium, mostly network computer manufacturers—many of them the founding members of PC/SC as well, wanted to define a framework that would support the smart cards on additional platforms such as network computers, smart phones, or set-top boxes [4]. As mentioned before, the OCF was tested on WinNT, Win95, Linux, IBM AIX, and several network computers. Obviously, the OCF and the PC/SC will coexist on Wintel platforms. Actually, the present OCF implementation can utilize PC/SC on the Wintel platform (see Section 12.6.6).

12.6.2 Operating System

The PC/SC ICC resource manager (see Section 12.5) is intended to be integrated within the operating system on the application terminal. In future versions of Microsoft Windows and Windows CE it will be an integral part of the operating system. Because it is written in Java, the OCF is independent of the operating system.

12.6.3 Terminal Application

The OCF addresses transparency for the application programmer with regard to card operating systems, card terminals, and card issuers. The PC/SC is primarily aimed at transparency with regard to card terminals and to card operating systems (to some extent), but not to the card issuer.

The PC/SC Version 1.0 does not provide multiapplication support; this has, however, been announced for Version 2.0. The OCF already provides functions for installation, removal, enumeration, selection of applications on the card, and name resolution for data files on the card (mapping user-friendly names to ISO 7816-4 file references). This functionality is offered in the form of a card management component. In other words, the application developer can program the application on the basis of an abstract file system model, without having to be concerned about where on the card a card issuer will install the application.

12.6.4 Programming Language

The PC/SC offers high-level C or C++ application APIs, but could in principle provide APIs in other programming languages as well. The OCF was created for the Java programming environment. One of the most popular Java features is "write-once/run-everywhere." For example, the OCF mechanisms for adapting the framework to a particular card operating system, card terminal, or card issuer allow for downloading the missing components for a particular card from the Internet, and adding them to the framework.

12.6.5 Architecture

This section gives a simplified top-down comparison of the OCF and the PC/SC Version 1.0 architectures [10]. For better understanding, please refer to Figure 12.5 as you read this section. For a thorough understanding, the reader may want to refer to Chapter 13, which provides explanations of the OCF components mentioned in this section.

The application-specific services and application management services are offered by the ICC service providers in the PC/SC, and by the Card-Services in the OCF. In contrast to the PC/SC, however, the OCF enforces a clear separation between the application-specific services and the application management services (i.e., between the CardService and ApplicationManagerCardService). Another difference is in the approaches to cryptographic services. The PC/SC defines a generic crypto interface through the cryptographic service provider. The OCF offers an optional package, opencard.opt.security, that provides cryptographic functionality. Furthermore, there is no corresponding part in the PC/SC for the OCF CredentialStore and CredentialBag.

The OCF layer containing the CardService is organized differently from the PC/SC service provider layer. For example, there is no directly corresponding part in the PC/SC for the OCF CardService.

The general structure of the resource manager, which is a central element of the PC/SC architecture, is quite different from the OCF architecture. The functionality of the OCF CardServiceFactory is covered by a part of the PC/SC ICC resource manager. The OCF layer comprising the CardServiceScheduler and the CardID roughly corresponds to a part

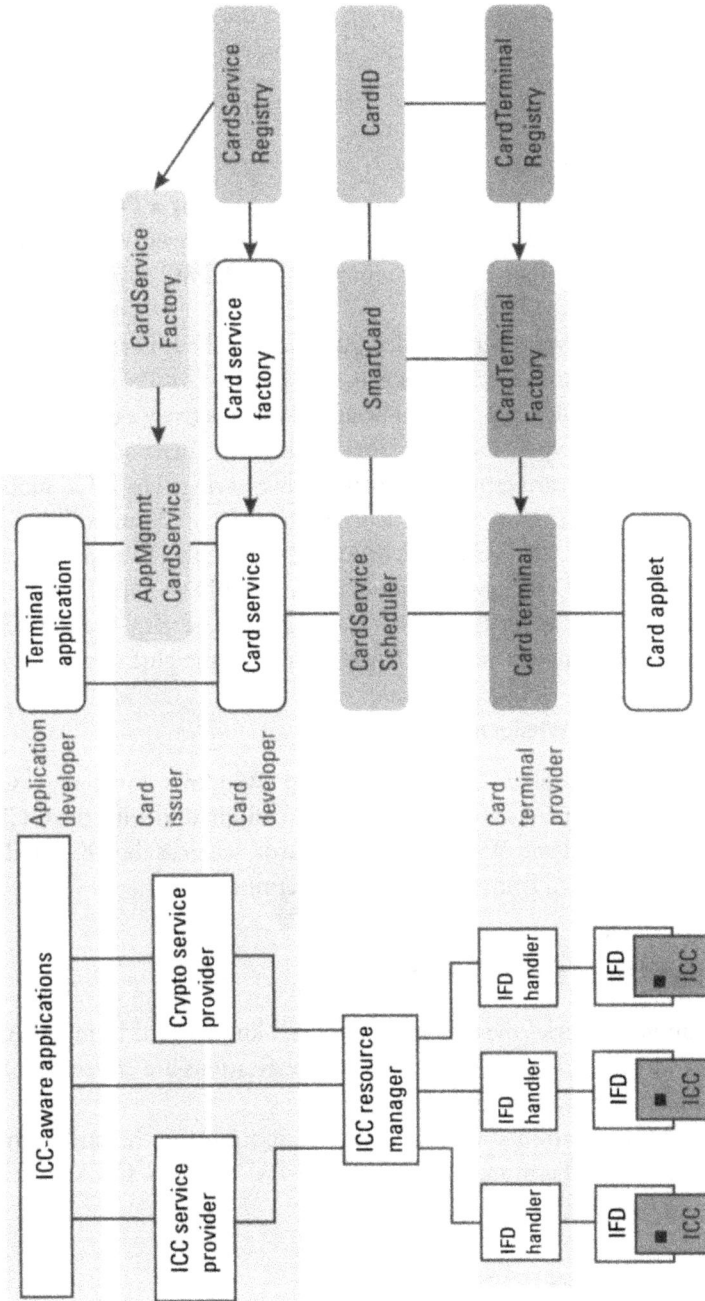

Figure 12.5 OCF versus PC/SC comparison.

of the functionality implemented by the PC/SC resource manager. The OCF `SmartCard` object has no corresponding part in the PC/SC.

Finally, the OCF `CardTerminal` has a functionality similar to that of the ICC IFD subsystem.

12.6.6 Interoperability

Within the Windows environment, the OCF can deploy a PC/SC card terminal driver by using the PC/SC ICC resource manager. A PC/SC card terminal already exists that takes advantage of the PC/SC resource manager and can be used for the OCF.

The PC/SC provides very detailed guidance for the IFD subsystem. The OCF, on the other hand, specifies only an API for the `CardTerminal` components. In this the OCF support can be added in a relatively easy way for different card terminal types, including smart tokens or smart accessories. The number of supported card terminals is constantly growing. The OCF supports the existing PC/SC IFDs as the OCF `CardTerminal`. Specifically, it was possible to create an OCF class, `PCSC10CardTerminal`, that can use a PC/SC-compliant IFD subsystem (consisting of an IFD and an IFD handler) on Wintel platforms as an OCF `CardTerminal` (see also Section 13.4), although this unfortunately does not always work in a straightforward way.

12.6.7 Java Card and Windows for Smart Card

Both Java Card–compliant smart cards and Windows for Smart Card (WfSC)–compliant smart cards can be used in conjunction with the OCF or the PC/SC. Java Card and WfSC specify the cards, whereas the OCF and the PC/SC "drive" the card from the application terminal.

12.7 Other Card Terminal APIs

This section gives a brief overview of some well-known card terminal APIs. In contrast to the OCF and PC/SC, these specifications are concerned with transparent access to card terminals only.

Europay International[5] was the first to concentrate on the card terminal software in its Open Terminal Architecture (OTA)[6] in 1996. OTA originally

---

5. http://www.europay.com.
6. http://www.europay.com/smartcard/Smartcard_ota_page.html.

included a Forth virtual machine that was extended to be language neutral so that code could be compiled from languages other than Forth. Currently, the C programming language is primarily used. In February 1999 Europay submitted the OTA specification for ISO standardization (ISO/IEC JTC1) in the form of a publicly available specification (PAS). Any organization that develops and owns PASs that are considered useful for conversion into international standards is eligible to apply for PAS recognition. The way in which the organization develops and maintains the specifications has to meet certain criteria.

The *Card Terminal Application Programming Interface* (CT-API) is a German interindustry standard for card terminals [12]. Its authors are Deutsche Telekom AG, GMD Forschungszentrum Informationstechnik GmbH, TÜV Informationstechnik GmbH, and TeleTrust Deutschland. It specifies the application-independent card terminal functions required to simplify the handling of and communication with smart cards in such a way that both memory and processor cards can be used, that the commands to control the card terminal can be transmitted, and that access to the card terminals can be independent of the port. A CT-API function library is provided by the card terminal vendor for the systems that are intended to be supported.

The Small Terminal Interoperability Platform (STIP) Consortium is also developing a standard for smart card terminals. The consortium was initiated by the Java Card Forum in 1999. The main goal is to specify a software platform that can be implemented on small devices with limited resources (e.g., card terminals, vending machines, POS terminals, magnetic-strip readers) and that provides support for multiple secure transaction applications on a terminal as well as interoperability of applications with different device types [13]. In addition, platform and application life cycle management should be provided. STIP Version 1.0 specifies a basic application framework and device access APIs to allow development of portable applications. As of February 2001, only a preview of Version 2 is available; it announces some API extensions and new APIs (e.g., cryptography, multiple-application support).

Visa International's Open Platform addresses both the smart card and the card terminal (see also Section 9.3). The Open Platform Terminal Specification divides the terminal component into two parts: a component for interaction with the smart card and a card-terminal-dependent part customized to a specific service provider's needs. The interoperability of the operation as a whole is not addressed.

References

[1] Hansmann, U., et al., *Smart Card Application Development Using Java*, Berlin: Springer-Verlag, 2000.

[2] Muller, C., "OpenCard Introduction & Overview. A Framework Approach to Smart Cards Services," Gemplus Research Lab, 2000; available at http://www.gemplus.fr/developers/technologies/opencard/.

[3] Schaeck, T., and P. Bendel, "OCF for Embedded Devices," RFC-15, Aug. 1999; available at http://www.opencard.org/work/rfc-15/rfc.15.txt.

[4] Hassler, V., and O. Fodor, "OpenCard Framework Application Development," *Dr. Dobb's Journal,* Vol. 25, No. 2, 2000, pp. 70–76.

[5] Gamma, E., et al., *Design Patterns CD. Elements of Reusable Object-Oriented Software*, Reading, MA: Addison-Wesley, 1998.

[6] Nemirovsky, A. M., "Building Object-Oriented Frameworks," Taligent White Paper, Dec. 1997; available at http://www-106.ibm.com/developerworks/library/oobuilding/index.html.

[7] PC/SC Workgroup, "PC/SC Workgroup Specifications 1.0 Download," 2000; available at http://www.pcscworkgroup.com/Specifications/SpecificationsDownload.html.

[8] Corcoran, D., "M.U.S.C.L.E.: Porting the PC/SC Architecture to Linux," presented at Gemplus Developer's Conference '99, Paris, June 21–22, 1999.

[9] PC/SC Workgroup, "Presentation of the Interoperability Specification for ICCs and Personal Computer Systems, Revision 2.0," White Paper, 2000; available at http://www.pcscworkgroup.com/Specifications/SpecificationsUpdate1.html.

[10] Seliger, F., "OpenCard and PC/SC—Two New Industry Initiatives for Smart Cards," 1999; available at http://www.opencard.org/docs/ocfpcsc.pdf.

[11] OpenCard Consortium, "OpenCard Framework (OCF): Frequently Asked Questions: Positioning of OCF Versus PS/SC"; available at http://www.opencard.org/misc/OCF-FAQ.shtml-PCSC.

[12] Deutsche Telekom AG, et al., "CT-API 1.1: Application Independent Card Terminal Application Programming Interface for ICC Applications," 1996; available at http://www.microdatec.de/download/ctapi11e.pdf.

[13] Small Terminal Interoperability Platform Consortium, "STIP Specification 1.0. Overview Version 1.0," 2000; available at http://www.stipgroup.org/.

13

OCF Structure

This chapter explains the OCF structure in more detail, in particular the card terminal layer and the card service layer. The core OCF Java packages can be divided into three main groups:

1. The utility packages contain support classes used throughout the framework.
2. The card terminal packages manage access to a card terminal (Section 13.1).
3. The card service packages provide the necessary infrastructure to interact with a card operating system and an on-card application (Section 13.2).

This chapter also discusses security aspects (Section 13.3) and communication with the card terminal as a physical device (Section 13.4). Finally, Section 13.5 explains how the OCF works with Java Card applets.

Figure 13.1 shows the OCF and Java packages mentioned in this chapter. The purpose of the chapter is not to provide a complete OCF reference manual, but to give explanations of the basic OCF functionality. The core packages are mandatory for all OCF implementations. The optional packages might not be available in all OCF implementations. For simplicity, the methods are generally given without arguments, and the exceptions are not discussed. The last part of this book will provide a simple example and a programming guide to illustrate how to implement an OCF card service. The

Card service layer	Standard card service interfaces:
	`opencard.opt.applet.mgmt`
	
	Card service:
	`opencard.core.service`
	`opencard.opt.service`
	
	Security:
	`opencard.opt.security`
	`java.security`
	`java.security.interfaces`

Card terminal layer	Card terminal:
	`opencard.core.terminal`
	`opencard.core.event`
	`opencard.opt.terminal`
	`opencard.opt.terminal.protocol`
	
	OCF card terminal communication:
	`javax.comm`
	`com.ibm.opencard.terminal.pcsc10`

Card applet layer	Java Card applets: `opencard.opt.applet`

Figure 13.1 OCF and Java packages mentioned in this chapter.

reader may refer to [1–3] for a detailed explanation of the OCF classes and mechanisms.

13.1 OCF Card Terminal Layer

OCF allows both static and dynamic configuration of card terminals. With the static configuration, the card terminals are known at system startup (e.g., serial port). With the dynamic configuration, additional terminals can be added at run time (e.g., PC/SC, PC card). A new card terminal can be added to the framework by implementing a CardTerminal, that is a device driver implementing the device-specific code, and a CardTerminalFactory, which the OCF uses to create an instance of the CardTerminal. A terminal application does not usually see the CardTerminal layer unless it wants to access some special features, such as reader display and keyboard management. The CardTerminal layer encapsulates the following components:

- Classes that allow access to a card terminal;

- Interfaces that can be implemented by the card terminal;

- Exceptions to be thrown at errors.

The components that provide the core functionality are organized in the `opencard.core.terminal` and `opencard.core.event` packages. The components that provide extended device support are contained in two packages, `opencard.opt.terminal` and `opencard.opt.terminal.protocol`. The following two sections describe these four packages.

13.1.1 Core Part of the Card Terminal Layer

The package modeling the terminal (`opencard.core.terminal`) contains classes whose instances represent physical devices and their characteristics. The main device abstractions are the `CardTerminal` class and the `CardID` class.

The `CardID` class represents a smart card's ATR response (see Section 5.1), which is used to identify cards. A subset of the ATR consists of what are known as historical characters, which may help determine the card type or even the applications supported by the card. This object can also contain the identification number of the slot containing the card.

Concrete implementations of specific terminal drivers (e.g., GemplusSerialCardTerminal) derive from the `CardTerminal` abstract class. By invoking the `CardTerminal` methods, a terminal application can retrieve information about a card terminal attached to the system, such as device name (`getName()`), device type (`getType()`), device address (`getAddress()`), or the number of slots (`getSlots()`).

A card terminal (e.g., a card reader) may have one or more slots, each accepting one card. A physical slot is modeled by the `Slot` class. A smart card inserted into a specific slot can be accessed through its `SlotChannel` object. A slot channel is exclusive because only one object can gain access to the corresponding smart card at a time. It is used to interact with the card, for example, to send and receive APDUs (`sendAPDU()` returns a `ResponseAPDU`).

The `CardTerminal` class provides methods for dealing with slot channels, such as opening and closing a slot channel (`openSlotChannel()`, `closeSlotChannel()`), checking whether a slot channel is available (`isSlotChannelAvailable()`), or sending and receiving APDUs through a slot channel.

As explained in Section 12.4.3, there is only one instance of the `Card-TerminalRegistry` class, whose role is to maintain a list of the installed `CardTerminals` registered within a system. When the `CardTerminal-Factory` creates a new `CardTerminal` instance, it registers the terminal with the `CardTerminalRegistry` using the `add()` method. An interested party can implement the `Observer` interface that is used by the `CardTerminalRegistry` to communicate added/removed card terminals and card insertion/removal. This interface decouples the `open-card.core.event` package, which provides the `CTListener` interface with a similar functionality (explained later), from the `opencard.core.terminal` package and in this way allows use of the terminal package without the events package in resource-constrained environments.

The `CommandAPDU` class and the `ResponseAPDU` class model the APDU transport packets. A `ResponseAPDU` represents an APDU received from the smart card in response to a previous `CommandAPDU`. They both inherit from the `APDU` class representing an APDU, which is the basic unit of communication with a smart card.

For some applications, such as electronic purses or digital signatures, only a legitimate user may be allowed to invoke an on-card service. This process is called cardholder verification (CHV) and, in practice, it means that the user is expected to provide some verification information, such as a PIN or a password (CHV information). For such card services the CHV information must be added to the command APDU that is sent to the card. The `CHVControl` class is a container class for CHV characteristics such as password encoding. An object of this type is usually created by the `Card-Service` and passed to the `CardChannel` by use of the `sendVeri-fiedAPDU()` method (see also Section 13.2.1.4).

With a simple card terminal, a PIN is typed into the application terminal (i.e., a PC), but in this case there is a possibility of eavesdropping on the PIN if the application terminal is not trusted. Better security is offered by card terminals with a display and a PIN pad. The card terminal can prompt the user on the display to enter his PIN by using the PIN pad. Such a `Card-Terminal` may implement the `VerifiedAPDUInterface` to query the user for CHV information and insert it into the `CommandAPDU`.

Considerable differences may exist, however, in how card terminals with extended functionality are programmed. A `CardTerminal` programmer can develop support for such a terminal. The `CardTerminalIO-Control` class contains parameters related to the card terminal's PIN pad and display, such as the supported character set. The `CardTerminalIO-Blender` is an abstract class that can be implemented to support special requirements such as data formatting. An implementation of this class can be provided by the `CardService`.

To detect the insertion or removal of a smart card, a card terminal provider essentially has two options: She may use the `EventGenerator`'s polling thread or provide her own. The first possibility is achieved by letting the implementation of the `CardTerminal` class implement the `Pollable` interface. All pollable terminals should call the `CardTerminalRegistry`'s `addPollable()` method in their `open()` method. The interval at which the `EventGenerator` polls the registered pollable card terminals can be changed by calling its `setPollInterval()` method.

The `EventGenerator` (from the `opencard.core.event` package) is a singleton object (see Section 12.4.3) that monitors the state of the card terminals attached to the system. This is a significant difference from OCF Version 1.1, in which events used to be generated by the `CardTerminalRegistry`. If a card is inserted or removed, the `EventGenerator` dispatches the `CardTerminalEvents` to the interested parties that have registered as listeners. The events can be passed on through the `CTListener` interface methods, `cardInserted()` and `cardRemoved()`.

13.1.2 Optional Part of the Card Terminal Layer

The optional part of the card terminal layer is provided within two packages, as mentioned in the introduction. The `opencard.opt.terminal` package extends the functionality of the core terminal package in the following way:

- The `ISOCommandAPDU` class adds full support of the ISO 7816 standard (see also Section 1.5), which means that all seven APDU cases are supported.

- The `AbstractLockableTerminal` class serves as a base for implementing lockable `CardTerminals`. In addition, `Lockable` is a generic interface for locking a terminal or individual slots.

- To be able to power up and down a smart card in a slot, a `CardTerminal` may implement the `PowerMangementInterface`.

- The `UserInteraction` interface makes it possible to access the display and the PIN pad of a card terminal directly. For example, information can be displayed to the user through the `display()` method, or retrieved from the user through the `keyboardInput()` method.

- The `TerminalCommand` is a generic interface for sending commands to a `CardTerminal`.

The `opencard.opt.terminal.protocol` package is interesting for the card terminal driver developer. It does not define an application interface. The current implementation provides support for a limited ISO 7816 T = 1 protocol variant.

13.2   OCF Card Service Layer

A developer of an off-card application need not know the smart card details, but only the high-level interface offered by a specific `CardService`. In other words, the application developer can write his own `CardService` to encapsulate smart-card-specific details. This is very useful if the terminal application is supposed to support smart cards of different types.

For some of the standard card service interfaces, the implementations should be provided by card issuers (Section 13.2.3). An implementation of a `CardService` can use the core components (the `opencard.core.service` package) or the optional components (the `opencard.opt.service` package) of the OCF card service layer. The following sections describe the core and the optional components, as well as the currently available standard service interfaces.

13.2.1   Core Part of the Card Service Layer

The functionality of the core components of the `CardService` layer can be divided into five groups: (1) application access, (2) card access, (3) `CardService` support, (4) CHV support, and (5) exceptions [1]. The following sections give brief explanations of the first four groups; the exceptions are beyond the scope of this chapter.

13.2.1.1   Application Access

The application access group contains the `SmartCard` class, the `CardRequest` class, and the `CardIDFilter` interface. A terminal application uses the `SmartCard` class to access the OCF. This class encompasses the following methods:

- The static methods for the OCF initialization and shutdown (i.e., `start()`, which loads the configuration data into the system properties, and `shutdown()`);

- The instance methods used when a `SmartCard` object is obtained for an inserted card (e.g., `getCardID()`, `getCardService()`);

- The methods to obtain a `SmartCard` object (e.g., `waitForCard()`); another possibility to obtain a `SmartCard` object is to wait for a `CardTerminalEvent`. For this, the application has to register as a `CTListener` with the `EventGenerator` (see also Section 13.1.1) and pass the `CardTerminalEvent` object to the static `getSmartCard()` method to obtain a `SmartCard` object.

The `SmartCard` class makes the functionality of many other OCF classes (e.g., `CardServiceRegistry`) available to the terminal application developer. This approach reduces the complexity of the framework because the developer does not have to know the details of these other classes [1].

The `CardRequest` class includes methods for setting and querying the smart card characteristics (e.g., `setCardTerminal()`, `getCardTerminal()`) and sets some important constants determining the behavior when the `waitForCard()` method is called. The default behavior is that a card already inserted is detected (ANYCARD), but it can be changed so that the terminal application waits for a new card insertion (NEWCARD). Finally, the `CardIDFilter` interface, which is based on filtering the `CardID`, is implemented by a terminal application if it is necessary to wait for a card with a specific ATR (see Section 13.1.1).

13.2.1.2 Card Access

The card access group defines some classes that are relevant to the `CardService` programmers.

Like the `CardTerminalRegistry`, the `CardServiceRegistry` is a singleton because there may be only one instance of it in a system (see Section 12.4.3). Its task is to keep track of the installed `CardServiceFactory` objects and to use the factories to create `CardService` instances (see also Section 12.4.2). A reference to the `CardServiceRegistry` object can be obtained by calling the static `getRegistry()` method. An enumeration of the installed `CardServiceFactory` objects is returned by the `getCardServiceFactories()` method.

The `CardService` objects use the `CardServiceScheduler` and the `CardChannel` for accessing the smart card. The `CardChannel` is a logical communication channel to the card. Smart cards allowing multiple logical channels are not supported, so only one CardChannel object can be associated with a card. This class also provides methods for data communication with the card, such as sendCommandAPDU() or sendVerifiedAPDU(), including a user password or a PIN, which offers lower security than secure messaging (see Section 13.3). It is therefore advisable to subclass the CardChannel to implement additional security [1].

The CardServiceScheduler serializes access to the CardChannel if it is used by several card services. The caller may choose to wait for a channel to become available by setting the blocking parameter in the allocateCardChannel() method.

13.2.1.3   Card Service Support

The abstract CardService base class is provided by the opencard.core.service package. To implement specific functionality, this abstract base class must be subclassed. The CardService subclass offers a high-level interface to the terminal application and to other CardService subclasses.

The CardService subclass is instantiated by the CardServiceFactory by invoking the newInstance() method. After instantiation, the CardServiceFactory calls the CardService object's initialize() method. For this method the CardServiceScheduler and SmartCard objects are passed.

The abstract CardServiceFactory base class must be subclassed and extended to be able to create specific CardService objects. Its methods are protected so that they cannot be used from within the application. Of all the CardServiceFactories available for a particular smart card, one must be the primary CardServiceFactory. This is accomplished by letting only one CardServiceFactory per card implement the PrimaryCardServiceFactory interface.

Basically, the CardServiceRegistry calls the factory's getCardType() method, passing it a CardID and CardServiceScheduler object. The factory analyzes the CardID to find out whether it recognizes the card; for further information, the factory can use the scheduler to communicate with the card. The result from this analysis is stored in a CardType object. If the CardServiceRegistry wishes to determine which card services are supported for a specific card, it calls the factory's getClasses() method with the CardType object as a parameter. Finally, to instantiate a new CardService object, the registry calls the factory's getCardServiceInstance() method with all the necessary parameters.

13.2.1.4 Cardholder Verification Support

As mentioned in Section 13.1.1, the CardService creates a CHVControl object containing a user's password or PIN and passes it to the Card-Channel by using the sendVerifiedAPDU() method. The CardChannel obtains the CHV information either from the CardTerminal (see Section 13.1.1) or by using the CHVDialog interface, which is one of the core components of the card service layer. This interface provides only one method, getCHV(), whose only parameter is the CHV identification number, since a card may have more than one CHV value. A default implementation is provided by the OCF (DefaultCHVDialog) and uses a Java Abstract Window Toolkit (AWT)–based dialog box. A terminal application that desires to provide its own version may implement the CHVDialog interface and pass the class to the CardService's method setCHV-Dialog().

13.2.2 Optional Part of the Card Service Layer

The optional components extend the core part of the card service layer. The CardServiceInterface represents an interface to the public methods in a CardService. The OCF11CardServiceFactory class can instantiate CardServices for a specific smart card. As its name says, it provides compatibility with the OCF Version 1.1.

13.2.3 Standard Card Service Interfaces

The OCF Version 1.2 reference implementation currently provides three standard CardService interfaces. Definition of such services is a responsibility of the OpenCard Consortium (see Section 12.3), which also considers proposals for new standard interfaces. When a card application developer implements a standard CardService interface, the corresponding card can immediately be used in the existing OCF applications.

FileAccessCardService is an abstraction for dealing with files on a smart card. Its high-level interface is similar to Java's java.io package, so the programmer does not have to know the underlying standards. The opencard.opt.iso.fs package extends the core part of the card service layer with methods for accessing transparent and structured files on the card based on the ISO 7816 standard (see Section 1.5). In addition, this package extends the opencard.opt.security package with support for secure messaging (see Section 3.4).

One of the most significant applications of smart cards is the provision of digital signatures. A smart card provides secure storage for private keys and a secure tamper-proof computing platform for computing digital signatures (see, e.g., RSA in Section 2.2). The public key pair may be generated on the card, which is the preferred way, or imported from a key generation device (in a secure way). The digital signature function is triggered by entering the correct PIN. (In the future, some biometric methods may be added as well.) `SignatureCardService` is an abstraction for key import and validation (`KeyImportCardService`), generation of public key pairs and export of public keys (`KeyGenerationCardService`), and for generation and verification of digital signatures based on a public key algorithm (`Signature-CardService`).

Smart cards can host multiple applications. A smart card–aware terminal application should be offered a way to find out which card applets are available on the card and provide this list to the user to select from. This is accomplished by `AppletAccessCardService`. In addition, some applet management tasks that should be implemented on the card, such as creation, registration, or deletion of applets, are offered by `Applet-ManagerCardService`. For these services it is necessary for the card issuers to provide some meta information on the card that can be accessed from the card-aware terminal applications. Both interfaces can be found in the `opencard.opt.applet.mgmt` package.

13.3 OCF Security

Some aspects of security have already been mentioned in previous sections, such as cardholder verification or the standard card service interface for digital signatures (`SignatureCardService`). This section deals with the components provided by the `opencard.opt.security` package, which builds on the `java.security.interfaces` and `java.security` packages. Because of U.S. export control regulations, the exportable version of the OCF provides no implementations of cryptographic algorithms. The cryptographic functionality can be provided by the terminal application or by a specific `CardService` implementation.

OCF security functionality can be divided into four groups: (1) credentials, (2) cryptographic keys, (3) smart card keys, and (4) `CardService` interfaces [1]. The credential classes contain interfaces to cryptographic algorithms and keys. The `RSASignCredential` class uses an RSA key to compute a digital signature over a block of data. (It can also use the Chinese remainder theorem, CRT.) It is not freely exportable from the United States.

The DSASignCredential class has the same functionality, but uses the DSA algorithm. It is freely exportable because it uses the java.security.Signature class and does not implement the algorithm. The DESSignCredential class for single-block DES encryption can be found in the com.ibm.opencard.access package. All three classes implement the SignCredential interface. The PKACredential and SymmetricCredential are tag interfaces that allow card services or other credential users to distinguish public key algorithm (PKA) credentials from symmetric credentials.

The cryptographic key classes store symmetric, public and private keys for DES, DSA, RSA, and RSA with CRT. They all implement the base class methods from the java.security.Key interface. For example, the RSAPublicKey class implements the java.security.PrivateKey interface. The OCF DSA key classes additionally implement the DSAKey interface from the java.security.interfaces package.

Different cards use different ways to identify a key on the card, for example, file names, numbers, or logical names. When a card service interface needs to identify a key on a card, the PrivateKeyRef or the PublicKeyRef interface should be implemented. For example, the PrivateKeyFile and PublicKeyFile classes implement these interfaces for smart cards with an ISO file system.

The CardService interface classes make secure messaging possible between the terminal application and the smart card (see Section 3.4). Basically, a terminal application that wants to access protected data on a smart card must authenticate itself to the card. This is done in such a way that the application presents its credentials to the card service, which in turns sends the credentials to the card. However, card services supporting a particular smart card type or smart card family require specific credentials and a credential store to put the credentials into. The CredentialStore abstract class is the base class for all stores. Therefore, the terminal application puts all credentials for a specific card family into a CredentialStore, and then all credential stores for all supported card families into a CredentialBag. The CredentialBag is supplied to the card service so that the application does not have to be concerned with which card it is currently communicating with. The bag is passed by using the SecureService.provideCredentials() method, which accepts an additional parameter defining the card area for which the bag is valid (SecurityDomain), for example, a path to the directory in which the on-card application resides. To support secure messaging, a CardService must implement the SecureService interface.

The `Credential` tag interface can be extended by a `CardService` subclass to support additional cryptographic algorithms (i.e., those not already supported by the OCF).

Finally, there are two more `CardService` interfaces: `CHVCardService` for cardholder verification and `AutCardService` for challenge-response-based internal authentication (when the smart card is authenticated) and external authentication (when the application is authenticated).

13.4   OCF Card Terminal Communication

An implementation of the `CardTerminal` represents a device driver for a card terminal that must somehow have the possibility to communicate with the physical device. In general, three possibilities are available for communicating with a card terminal; they will be briefly described in this section (see also Figure 13.2).

The first possibility is to use the Java Communications API [4], also referred to as `javax.comm.`, which is the name of the extension package. It provides a platform-independent method for accessing serial and parallel ports from Java programs. Card terminals of many types are attached to serial ports, so `javax.comm` can be used to program a `CardTerminal`. The `javax.comm` interfaces provide method calls to list the available communication ports, to control shared and exclusive access to ports, and to access and change the port parameters, such as baud rate, number of stop bits, or parity generation. The main high-level abstractions are the `CommPort` and the

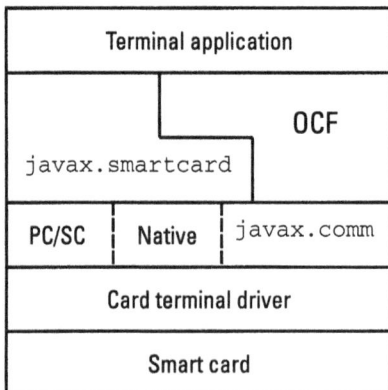

Figure 13.2   OCF card terminal communication.

`CommPortIdentifier` class for managing access and ownership of communication ports. The low-level abstractions such as `SerialPort` and `ParallelPort` provide an interface to physical communication ports. The current `javax.comm` release allows access to serial (RS-232) and parallel (IEEE 1284) ports. A `CardTerminal` may register to receive events describing communication port state changes. For example, when a serial port has a state change, the `SerialPort` object propagates a `Serial-PortEvent` that describes the state change.

In cases where a card terminal does not use the standard serial or parallel ports or where `javax.comm` is not available for a particular platform, a card terminal driver can possibly be written that uses the Java Native Interface (JNI) [5]. The JNI enables a Java application to access operating system native binaries. The implementation consists of a Java class used by the Java application and a piece of code written in another programming language running in the operating system environment. The resulting code is, of course, platform specific and therefore not portable.

Finally, the third possibility is to run the OCF over the PC/SC (see Section 12.5), as mentioned in Section 12.6. The OCF reference implementation provides PC/SC `CardTerminal` drivers, such as the `Pcsc10Card-Terminal` class from the `com.ibm.opencard.terminal.pcsc10` package. On a Windows platform, the PC/SC card terminals (i.e., IFDs) can be accessed through the OCF if the OCF PC/SC `CardTerminal` is configured.

For the sake of completeness, we should also mention that the OCF can be used over the `javax.smartcard` package to replace a part of the OCF functionality (i.e., `CardTerminal`, `CardChannel`) [6]. This package provides abstractions for card terminal drivers (`IFDDriver`), card terminal slots (`Slot`), APDUs (`APDU`), card channels (`Channel`), and events (`SmartCardEvents`). It also provides interfaces for card terminals (`IFD`), its displays (`IFDDisplay`), and event and smart card listeners (`Channel-Listener`, `SmartCardListener`).

13.5 OCF and Java Card Applets

The use of Java Card applets within the OCF is based on a standard object-oriented design pattern called a *proxy* (sometimes also referred to as a *surrogate*). A proxy can be applied whenever there is a need for a more sophisticated reference to an object than a simple pointer [7]. Support for this mechanism is provided in the `opencard.opt.applet` package.

Each card applet is represented by an applet proxy on the platform on which the OCF is running [8]. This is a design pattern for a remote proxy that provides a local representative for an object in a different address space (in this case a smart card) [7]. Whenever the application invokes a proxy method, the proxy starts communication with the card applet on the card and generates some result, which it then returns to the application.

An application uses the card applet services through a corresponding card applet proxy. Consequently, a card proxy should be a subclass of the OCF CardService. The BasicAppletCardService class, from which the AppletProxy class is derived, extends the CardService. The AppletProxy contains an attribute that holds the application identifier (AppletID) of the card applet to which it is associated. The state of an applet can be represented through the AppletState to the associated proxies. The AppletState object also provides a means for the associated applets to cooperate.

An applet proxy may be instantiated several times by different threads. Concurrent access by different instances to the card is serialized by the CardServiceScheduler, as is usual for all OCF card services.

A single card can host multiple card applets. All AppletProxy instances associated with the same physical card share a common state object (CardState) to ensure a consistent view. The CardState object may be held by the CardChannel since there is only one card channel created per physical card.

The applet proxy mechanism is illustrated in Figure 13.3. Two card applets are available on a physical smart card, Card Applet 1 and Card Applet 2. Card Applet 1 is selected and Card Applet 2 is deselected, which is indicated by dashed arrows. All instances of Applet Proxy 1 can cooperate by using Applet State 1, which represents the state of Card Applet 1. All instances of Applet Proxy 1 and Applet Proxy 2 (i.e., all applet proxy instances associated with the same physical card) share a common state object Card State.

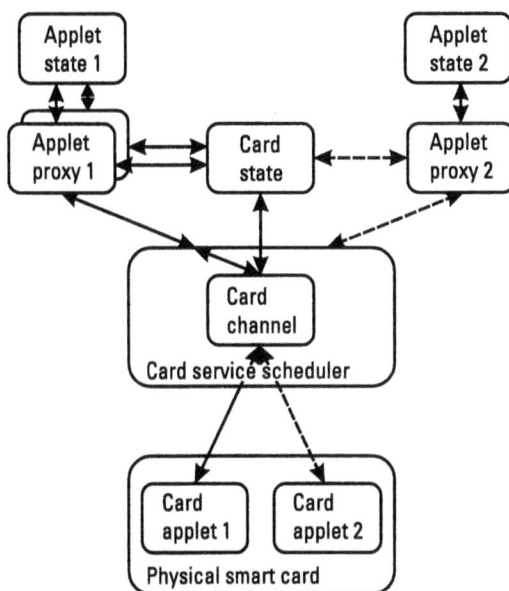

Figure 13.3 Applet proxies.

References

[1] Hansmann, U., et al., *Smart Card Application Development Using Java*, Berlin: Springer-Verlag, 2000.

[2] OpenCard Consortium, "API Docs V1.2"; available at http://www.opencard.org/docs/ 1.2/index.html.

[3] IBM, *OpenCard Framework 1.2 Programmer's Guide*, 4th ed., OpenCard Consortium, Dec. 1999; available at http://www.opencard.org/docs/pguide/PGuide.html.

[4] Sun Microsystems, "Java™ Communications API," 2001; available at http://java.sun. com/products/javacomm/index.html.

[5] Sun Microsystems, "JNI—Java Native Interface," 2001; available at http://java.sun.com/ products/jdk/1.1/docs/guide/jni/index.html.

[6] Sun Microsystems, "javax.smartcard Package Index," 2001; available at http://java.sun. com/products/commerce/docs/smartcard_api/packages. html.

[7] Gamma, E., et al., *Design Patterns CD. Elements of Reusable Object-Oriented Software*, Reading, MA: Addison-Wesley, 1998.

[8] Schaeck, T., and R. Di Giorgio, "How to Write OpenCard Card Services for Java Card Applets," *JavaWorld*, Oct. 1998; available at http://www.javaworld.com/ javaworld/jw-10-1998/jw-10-javadev_p.html.

Part IV
Case Study: Java Card Application Development with the OpenCard Framework

This part of the book is dedicated to the development of a sample EMV credit/debit application using Java Card technology and the OCF. It begins with an overall functional description of the application. From the architectural point of view, our sample application is split into two applications communicating with each other: a smart card application and a terminal application. We will demonstrate in one of the chapters how an EMV smart card application can be implemented on the Java Card platform in the form of a card applet. Further chapters will show how an EMV terminal application can be implemented using the OCF.

14

Case Study Overview

This chapter presents an overview of a sample credit and debit application that follows EMV specifications. It presents and discusses the architecture and functionality of the overall credit and debit application as well as the functionality of the terminal and smart card applications. Particular attention is paid to some EMV-specific aspects that are not supported in the sample application. This will help the reader to understand EMV credit and debit applications better and will fill the gap between the sample application and real-life EMV applications.

14.1 Sample Application Functionality

A general overview of EMV credit and debit applications was already presented in Chapter 7. One of the core ideas behind the sample application was to demonstrate how EMV applications can be developed in a manner that is independent of the platform and smart card manufacturer used, that is, by exploiting the benefits of the Java technologies, especially Java Card and the OCF. It was never our goal in this book to develop a complete and ready-to-use EMV application. Hence, the focus of the sample application is on the key aspects of EMV credit and debit applications.

14.1.1   Application Architecture

Basically, three core players are involved in a typical EMV application (see Figure 14.1). The first player is the smart card[1] itself. The card is responsible for performing a mutual authentication, approving a transaction initiated by the terminal, and making a range of decisions related to the transaction flow. All of these functions carried out by the card are discussed in this chapter.

The terminal is the initiator of a transaction. The operational environment of the terminal could be a POS, an ATM, or so forth. In our context, the word *terminal* stands not just for the mechanical device in charge of accepting the card, receiving entry of the PIN, and displaying text on its screen. We concentrate on the EMV-related functionality of the terminal and underlying applications, that is, the software.

On the one hand, the terminal is capable of initiating and carrying out the transaction with the card by itself. On the other hand, the terminal is also capable of making its own decisions influencing the transaction flow and establishing, if needed, an online session with the remote computer systems of the issuer.

The issuer is the banking institution that issues payment card products to individual users. The issuer's computer system may also be involved in a transaction if the terminal or the card decides to go online to carry out the transaction. The tasks of the online session with the issuer could be to perform an online authentication or to approve a transaction.

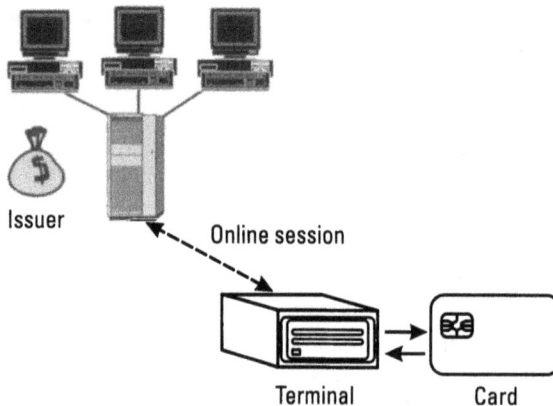

Figure14.1 General architecture of a typical EMV application.

---

1. Always referred to as the integrated circuit card in the EMV specifications.

For reasons of simplicity and space, we omit in our sample application all functions based on online communication to issuer computer systems. On the one hand, this decision results in certain rather significant limitations of the sample application functionality because some core security functions such as online authentication and online transaction approval become unavailable. On the other hand, it makes the sample application more transparent and easy to follow.

In the following sections of the chapter, we will always point out the implications of the defined limitations on the sample EMV application presented here.

14.1.2 Transaction Flow

The main EMV transaction steps were described in general in Part I (see Section 7.2). In our sample application, we stay with the limited number of transaction steps that are sequentially performed by the terminal and the card. Figure 14.2 shows a basic overview of the steps involved.

In the first step, the sample application must be selected on the card. The terminal accomplishes this task by sending to the card a SELECT command APDU containing the desired AID of the applet to be selected.

After the applet has been selected, the terminal retrieves processing options from the card. The processing options include data objects providing the terminal with information on the location of application-specific data on the card and also the transaction-specific functions supported by the card. A detailed look at these data objects as well as the definition of their content for our sample application can be found in Section 14.1.3.

In the next step, the terminal retrieves from the card a set of data objects referenced in the processing options obtained on the previous step. These data objects include various data related to the card application and conditions of its operation, as well as some information on the person who owns the card. The data objects are stored on the card in linear record files and therefore are retrieved by a sequence of READ RECORD commands. As soon as the terminal receives all necessary data objects from the card, it starts processing the information obtained from the card. This procedure is also referred to as restrictions processing. The goal of restrictions processing is to verify whether the use of the card and the transaction purpose are at all possible or allowed.

For reasons of simplicity, our sample application will provide no online card terminal authentication. Originally, the authentication, either static or

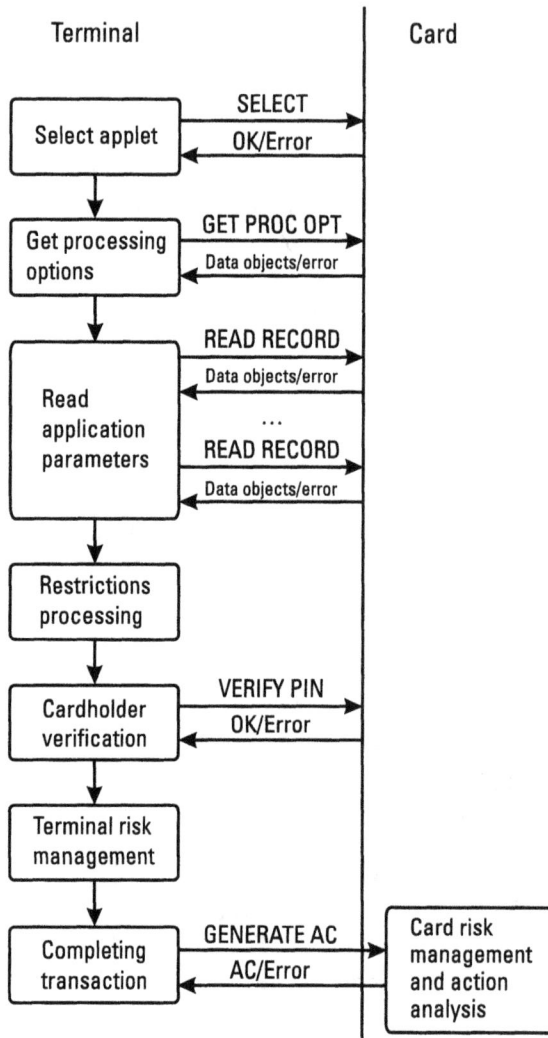

Figure14.2 Sample application transaction flow.

dynamic, must be performed after the retrieval of the processing options and
before restrictions processing.

Once the restrictions have been processed by the terminal, the next
transaction step is to perform cardholder verification. In other words, the
owner of the card must prove legitimate use of the card. Cardholder verification is based on verification of the PIN presented by the cardholder or on
the cardholder's handwritten signature, or on a combination of both. PIN

verification can be performed offline or online, where the PIN value is transmitted either encrypted or in plaintext. Because online functions are not supported in our sample application, we stay with offline PIN verification. Also, to keep the application simple, the PIN is to be transmitted from the terminal to the card in plaintext.[2]

After cardholder verification, the terminal proceeds to the terminal risk management. This procedure is performed by the terminal locally and aims to reach a decision on how the transaction is to be accomplished—declined, performed online, or performed offline. The fact that online functions are not supported in our sample application makes the application much simpler—the result of the terminal risk management procedure will be either to perform the transaction offline or to decline it.

After the terminal comes to a decision about how the transaction is to be accomplished, it requests the card to approve the decision. The request is sent to the card in the form of a GENERATE AC command APDU. On receiving the request, the card performs its own internal verifications of the transaction as well as the card risk management and card action analysis procedures. As a result, the card either approves the transaction or declines it. If the terminal has already declined the transaction, the card has no other choice than to approve this decision to decline.

Because the terminal never goes online and the transaction can thus only be performed offline, our sample application will support only the first GENERATE AC command (the second GENERATE AC command is used for completing transactions online). In this way, the sample application transaction is completed by the response to the first GENERATE AC command sent by the card to the terminal. Afterward, the terminal has only to verify the response.

All data structures used for the cryptogram calculation as well as the underlying algorithms will be presented in the following sections of the chapter.

14.1.3 Data Objects

Our sample application, like any real-life EMV credit/debit application, will deal with various data objects related to the application functions and transaction processing. Some of those data objects are generated by the terminal or by the card during the transaction processing, while some always reside

---

2. Otherwise, the PIN would have to be encrypted with an asymmetric encryption algorithm.

either at the terminal or on the card and are stored in either volatile or non-volatile memory. This section of the chapter gives a description of the core data objects defining our sample application that are used throughout the application.

The description starts with the data objects retrieved by the terminal after applet selection (see Figure 14.2).

14.1.3.1   Application Interchange Profile and Application File Locator

The AIP and AFL objects are received by the terminal in the response to the GET PROCESSING OPTIONS command APDU, the first command performed after the application selection.

AIP informs the terminal about the topmost functions supported by the card and is represented as a TLV object of the following format:

Tag (T)	Length (L)	Value (V)
82 H	2	2-byte value

According to the EMV specifications, the second (the rightmost) byte of AIP is reserved for future use and must be set to the value 00. For our sample application, we have coded the first (the leftmost) byte according to Table 14.1. In this way, the resulting value of our sample AIP will be 5C 00 H.

After reading this AIP, the terminal will know that the card supports cardholder verification and offline data authentication, that the terminal must perform the risk management, and, finally, that the issuer cannot be authenticated by means of the second GENERATE AC command.

The second object received by the terminal in response to the GET PROCESSING OPTIONS command is the AFL. The AFL contains information on AEFs maintained by the card application and specifies records in those files that the terminal should read in order to obtain all application- and cardholder-specific data. The AFL is composed of a sequence of 4-byte blocks, where each block in the sequence describes a particular elementary file. The AFL is coded as a TLV object of the following format:

Tag (T)	Length (L)	Value (V)
94 H	4 * *n*	Data on each from *n* files

Table 14.1

Sample AIP Coding of the First Byte

b8	b7	b6	b5	b4	b3	b2	b1	Meaning
0								This bit is not used.
	1							Offline data authentication is supported by the card.
		0						This bit is not used.
			1					Cardholder verification is supported by the card.
				1				Terminal risk management is to be performed.
					1			Issuer authentication is not supported via second GENERATE AC command.
						0		This bit is not used.
							0	This bit is not used.

As stated above, each elementary file is described in 4 bytes. Table 14.2 gives a description of AFL byte coding for a particular elementary file. If the AFL should contain data on more than one file, bytes 1–4 are repeated for each file as many times as files should be described in AFL.

Our sample application will have two elementary files. The first file will have SFI 1 and will contain six records, the second will have SFI 2 and will contain only two records. Hence, data on the first file is coded as

08 01 06 00 H

Table 14.2

AFL Byte Coding

Byte	Value
Byte 1	Bits 8–4: short file identifier (SFI), Bits 3–1: 0 0 0
Byte 2	First record number to read from this file
Byte 3	Last record number to read from this file
Byte 4	Number of consecutive records signed in signed application data

and on the second as

10 01 02 00 H.

The TLV object containing the sample application AFL will be defined as

94 08 08 01 06 00 10 01 02 00 H.

In concluding the description of AIPs and AFLs, we should mention that their content is defined and put on a card during the card personalization stage and is never changed afterward.

14.1.3.2   Application Data Objects

EMV specifications define a rather long list of application data objects, mandatory and optional, that must be maintained on the card and supplied to the terminal on request. In our sample application, we limit the list of application data objects to the following eight objects that will be stored in the sample application elementary files:

- *Application version number* (AVN): defines the version number assigned to the application by the payment system;

- *Application usage control* (AUC): contains issuer-specific restrictions on the geographic usage of the card application and also restrictions on payment services allowed for the card application;

- *Application effective date:* the date starting from which the card application can be used;

- *Application expiration date:* the date on which the card application expires and can no longer be used;

- *Issuer country code:* identifies the issuer's country according to ISO 3166 [1];

- *Application currency code:* identifies the currency according to ISO 4217 [2] in which the bank account related to the card application is managed;

- *CVM list:* defines a list of cardholder verification methods supported by the card;

- *Cardholder name:* defines the cardholder name according to ISO 7813 [3].

Whereas the meaning and the content of some objects from the list are easy to understand, others need a more detailed explanation.

Our explanation begins with the AUC object, which value states issuer-specific restrictions on the geographical usage of the card application and restrictions on the payment services supported by the card.

Geographical restrictions distinguish domestic and international payment operations. All payment services are grouped into four categories: cash transactions, goods, services, and ATMs. Each category can be either enabled or disabled for a particular card application. Because we do not wish our sample application to be subject to any kind of geographical or functional restrictions, we will enable all types of payment services for both domestic and international use. Our intention is reflected in the first byte of the 2-byte AUC object as shown in Table 14.3.

The value of the first byte of the sample AUC is defined as FF H. Bits of the second AUC byte are mainly reserved for future use. Only the last 2 bits of the second byte define whether cashbacks are allowed or not. We assume that the cashback function is not really relevant to our sample application and set the AUC's second byte value to 00 H. Hence, the sample AUC is defined as FF 00 H. Whereas our card application will support all types of payment operations worldwide, the terminal will be more restrictive and will allow payment for domestic goods only, that is, goods bought in the card's country of origin—Austria.

Another application data object that needs more detailed explanation is the CVM list. The CVM list is composed of at least 10 bytes and is extended by 2 bytes for each additional cardholder verification method supported by the card. The coding of the CVM list bytes is fairly complex and, in addition,

Table 14.3
Sample AUC Coding of the First Byte

b8	b7	b6	b5	b4	b3	b2	b1	Meaning
1								Valid for domestic cash transactions
	1							Valid for international cash transactions
		1						Valid for domestic goods
			1					Valid for international goods
				1				Valid for domestic services
					1			Valid for international services
						1		Valid at ATMs
							1	Valid at terminals other than ATMs

depends on the particular payment system to which the card application belongs. For reasons of space, we will omit a detailed explanation of the sample application CVM list coding and give just its value followed by a brief explanation of particular bytes. Following the Europay specification [4], we define the CVM list for our sample application as

00 00 00 00 00 00 00 00 41 03 H.

The last 2 bytes of the CVM list define the method code and conditions of method use, respectively. In our case, the sample application will use offline plaintext PIN verification (41 H), which must be used if the terminal supports this method (03 H). Hence, our sample card application will use the same CVM codes as any other MasterCard, Eurocheque, Cirrus, or Maestro application currently in use.

The meaning of the other application objects that will be supported on the sample application card is rather simple. Their values are given in Table 14.4, which summarizes descriptions of the application data objects.

Our sample application is to be effective starting January 1, 2000, and will expire on December 31, 2003 (03 01 1F H). The sample application issuer country code 00 28 H stands for Austria. The application currency is Euro (code 03 D2 H). Hans Mustermann, the John Doe of German-speaking countries, will be the honorable owner of the sample application card.

Table 14.4
Sample Application Data Objects

Object	Tag	Length	Value	Location (SFI/Rec. No)
Application version number	9F08 H	02	00 02 H	1/1
Application usage control	9F07 H	02	FF 00 H	1/2
Application effective date (YYMMDD)	5F25 H	03	00 01 01 H	1/3
Application expiration Date (YYMMDD)	5F24 H	03	03 0C 1F H	1/4
Issuer country code	5F28 H	02	00 28 H	1/5
CVM list	8E H	10	00 00 00 00 00 00 00 00 41 03 H	1/6
Cardholder name	5F20 H	15	"Hans Mustermann"	2/1
Application currency code	9F42 H	02	03 D2 H	2/2

The last column of Table 14.4 indicates the location of the data objects. The first number denotes the short file identifier of the elementary file; the second number denotes the record in that file.

14.1.3.3 Application Sequence Flag

The application sequence flag object is one of the key data objects maintained by the card application. Its fields reflect states of the card application throughout the transaction processing. In transaction steps that are critical for the transaction flow, the card application checks the object fields to determine whether all preconditions for performing the transaction step are satisfied.

To ensure physically that all fields of the application sequence flag object are cleared when the power is removed from the card, *the object must be located in the volatile memory of the card.*

Our sample card application will not implement all fields of the application sequence flag object that are defined in the EMV specifications. This is because the application will not implement all EMV application functions and thus many states will not be present in our card application at all. The application sequence flag object of the sample card application will contain the following fields:

- *Application selected:* Set when the application becomes selected.

- *Application is invalidated:* Set when the selected application becomes invalidated for certain reasons.

- *Get processing performed:* Set when the GET PROCESSING OPTIONS command has been performed.

- *PIN presentation performed:* Set when the VERIFY command has been performed.

- *PIN correctly verified:* Set when the PIN has been successfully verified, that is, that is, when the VERIFY command has been completed successfully.

- *AAC generated:* Set when the card responded with AAC to the first GENERATE AC command because the application is invalidated.

- *ARQC generated:* Set when the card responded with ARQC to the first GENERATE AC command.

In Chapter 15, we will demonstrate how the application sequence flag object can be implemented using transient (that is, cleared on card reset) data objects of Java Card.

14.1.4 Application Selection

Selecting the card application is the first transaction step that the terminal must accomplish. As already mentioned, every type of card application is distinguished by its AID. A card application is selected by means of the SELECT command APDU sent to the card, which contains the AID of the application to be selected.

Until now, our sample card application has had no name; in view of its purpose to demonstrate how EMV credit/debit applications can be developed using Java Card and OCF, we hereby define the card application AID as "EMVdemo" or as a binary string containing the characters' codes as follows:

$$45\ 4D\ 56\ 64\ 65\ 6D\ 6F\ H.$$

According to the EMV specifications, our card application should be selectable by the SELECT command APDU having the following format:

CLA	INS	P1	P2	L$_c$	Data	L$_e$
00	A4	04	00	AID length	AID	00

The AID length must be at least 5 and at most 16 bytes. Our card application AID "EMVdemo" fits this range perfectly.

In contrast to the EMV specifications, the response to the command in the case of successful application selection will consist only of the status word 90 00 H and will contain no FCI. This we do for simplicity's sake and because the FCI only contains some payment-system-specific parameters that have no use in our sample application.

If the application selection fails, the card will report the error by means of one of the following status words:

SW	Meaning
69 99 H	No application (applet) selected
6A 86 H	Wrong parameters P1 and P2
69 85 H	Conditions of use are not satisfied

The status word 69 99 H will be returned if no application with the given AID was found on the card. If the parameters P1 and P2 of the SELECT command APDU had an incorrect value (not as specified above), the card will return the status word 6A 86 H. And, finally, if the application cannot be selected because of certain security restrictions defined for the card, the card will answer with the status word 69 85 H.

In the case of successful selection, the application must reset all fields of the application sequence flag object (see previous section) except the field Application Selected, which is set to indicate successful application selection.

14.1.5 Processing Options Retrieval

If the card application was successfully selected, the terminal proceeds with retrieving the processing options. In fact, this action initiates a transaction with the card. In turn, the card verifies whether the application was selected.[3] This it does by checking the respective field of the application sequence flag object and rejecting the command if the object field is reset.

The processing options are retrieved by the terminal in a response to the GET PROCESSING OPTIONS command APDU that has the following format:

CLA	INS	P1	P2	L$_c$	Data	L$_e$
80	A8	00	00	02	PDOL data	00

The Data field of the command contains the PDOL. PDOL is used by the terminal to specify a desired AIP/AFL pair it needs to read from the card. In fact, a card application can manage not just one AIP/AFL pair, but several of them. Therefore, the card application can choose an appropriate one to send back to the terminal depending on the received PDOL.

Our EMVdemo application manages just one AIP/AFL pair. Therefore, it would expect to receive an empty PDOL because it can deliver to the terminal only that AIP and AFL. According to the EMV specifications, an empty PDOL is coded as a TLV object in the following form:

Tag	Length
83 H	00

---

3. Actually, such verification is not really necessary for Java Card—if the desired applet was not previously selected, the GET PROCESSING OPTIONS command APDU will not reach the applet and will either be rejected by JCRE, if no applet on the card is selected, or be passed to any other currently selected applet.

In processing the GET PROCESSING OPTIONS command, the card application must first check whether this command was already executed during the current transaction. This check is done using the Get Processing Performed field of the application sequence flag object, which must be reset. If the field is set, the command processing will be terminated and the card application will answer with the "Conditions of use are not satisfied" status word (69 85 H).

The card application maintains an internal counter that tallies all transactions processed by the card. This counter is called the application transaction counter (ATC). When the transactions start with the GET PROCESSING OPTIONS command, the ATC is incremented any time the command is successfully completed.

For any payment application, a maximum number of transactions that a card can perform is predefined by the card issuer. Therefore, each time that the GET PROCESSING OPTIONS command is processed, the card application checks whether the ATC has reached its limit. If the limit is reached, the application is assumed to be invalid and the card answers with the "Reference data invalidated" (69 84 H) status word.

For our EMVdemo application, we define the ATC limit to be FF FF H.

If the limit is reached, the card application will become invalidated forever. We define no means to reset the ATC value (usually this is accomplished by using issuer scripts).

If the checks of the ATC value and the application sequence flag object fields Get Processing Performed and Application Selected were successful, the card application will return the AIP and AFL to the terminal packed into a TLV object having the following format:

T	L	T	L	V	T	L	V
77 H	Object length	82 H	02	AIP	94 H	AFL length	AFL

However, before sending out the AIP and AFL, the card application must update the fields of the application sequence flag in the following way:

1. Set the Get Processing Performed field;
2. Reset the ARQC Generated and AAC Generated fields.

If the command processing fails, the card will report the error by one of the following status words:

SW	Meaning
67 00 H	Wrong length
6A 86 H	Wrong parameters P1 and P2
69 84 H	Reference data invalidated
69 85 H	Conditions of use are not satisfied

The card will answer with the "Wrong length" status word if a nonempty PDOL object is received, that is, if the length of the data field of the GET PROCESSING OPTIONS command APDU is not equal to 02.

14.1.6 Reading Application Parameters

As mentioned previously, the AFL object contains identifiers of application elementary files as well as the numbers of records containing various application-specific parameters that the terminal needs to fulfill a transaction. The terminal receives the AFL in response to the GET PROCESSING OPTIONS command.

In the next transaction step, the terminal simply retrieves record by record all of those parameters from the card application elementary files referenced in AFL. This is done using the READ RECORD command, which is performed as many times as there are records that the terminal should read from the card.

The READ RECORD command is simple and easy to process. The respective command APDU has the following format:

CLA	INS	P1	P2	L$_e$
00	B2	Rec. No.	RCP	00

Parameter P1 contains the record number to be read. According to the EMV specifications, the record number must be different from 0. The file from which the record is to be read is specified by the reference control parameter (RCP), whose value is contained in the parameter P2. Bits 8–4 of RCP contain the SFI of the desired file. Bits 3–1 must be set to "1 0 0" to indicate that parameter P1 contains the record number.

Checks performed by the card application while processing the READ RECORD command APDU are rather basic and cover parameters P1 and

P2. First of all, if parameter P1 is equal to 00, the card will answer with the status word "Function not supported" (6A 81 H). The card will also answer with the same status word if bits 1–3 of parameter P2 are not equal to "1 0 0."

If the file specified in P2 is not found, the card will answer with the status word "File not found" (6A 82 H) and, if the file contains no record specified in P1, the card will answer with the status word "Record not found" (6A 83 H).

A successful response to the command will contain the desired record followed by the success status word 90 00 H. A summary of error status words that may be returned in response to the READ RECORD command is given below:

SW	Meaning
6A 81 H	Function not supported
6A 82 H	File not found
6A 83 H	Record not found

After the terminal reads all file records from the card, it proceeds with the restrictions processing procedure (see Figure 14.2). In our sample application, the restrictions procedure is rather simple. It will therefore be explained directly in Chapter 16, where the terminal application is presented.

14.1.7 Cardholder Verification

Cardholder verification is one of the steps essential to the transaction. In this step, the person presenting a card for a certain payment operation must prove that she may legitimately use the card. In most of the payment applications of today, cardholder verification is based on a PIN verification, a handwritten signature verification, or a combination of both. The same methods are also used in EMV applications. As mentioned at the beginning of the chapter, we will use one of the simplest PIN verification methods—offline plaintext PIN verification. The word *plaintext* means in this context that the PIN is transmitted from the terminal to the card in unencrypted form.

Like any other payment application, our card application will maintain a PIN try counter reflecting the number of unsuccessful attempts to verify the PIN. If the PIN try counter reaches a predefined value (a try limit), the PIN will become blocked, which means that even presenting the right PIN value subsequently will not lead to successful PIN verification. The PIN can be unblocked only via a special command PIN UNBLOCK.

We would like to point out that it is very important for the PIN try counter to be incremented before each PIN verification and be reset if the verification is successful. Otherwise, an adversary could remove power from the card immediately after PIN verification and in this way prevent the PIN try counter from incrementing. If the PIN try counter were never incremented, the PIN would never become blocked. This would give the adversary a chance to perform a brute force attack, that is, in which he could try all possible values on the PIN in order to find its correct value.

To avoid complexity, our EMVdemo application will not support this command. This means that if the PIN becomes blocked, it will remain blocked forever.

We set the PIN try limit value to 3.

In other words, a cardholder will be allowed only three tries to present the PIN correctly.

To perform PIN verification, the terminal sends to the card the VERIFY command APDU, which has the following format:

CLA	INS	P1	P2	L$_c$	Data
00	20	00	80	08	PIN Data

The parameter's P2 value of 80 H denotes that the PIN data are transmitted in plaintext. PIN Data contains not only the PIN value but also some additional information and has a complex coding. The length of the PIN Data is always 8 bytes. The first byte contains two values—the control field placed in the leftmost digit of the byte and the PIN length placed in the rightmost digit of the byte. The other 6 bytes contain either the PIN value digit by digit (the PIN must have a minimum of four digits) or filler digits FF H. The last, the 8th byte of the PIN Data, is always filler digits FF H.

We will explain the coding of the PIN Data by an example. For our EMVdemo application, we define the PIN value to be 1234.

This value will be encoded in the PIN Data as follows:

$$24\ 12\ 34\ FF\ FF\ FF\ FF\ H$$

The first digit of the first byte is the control field, 2, the second digit of the byte is the PIN length, 4. The second byte contains the first two digits of the PIN, 12, the third byte contains the last two digits of the PIN, 34. All other bytes of the PIN Data are filler bytes FF H.

As with any other command processing, the card application must check values of the VERIFY command parameters P1 and P2. If they do not match predefined values (see above), the card will answer with the status word "Wrong parameters P1 and P2" (6A 86 H). If the length of the PIN Data differs from the predefined value 08, the card will answer with the status word "Wrong length" (67 00 H).

Before starting the PIN verification process, the card application must update fields of the application sequence flag object, that is, reset the field PIN Verified, and set the field PIN Performed in order to reflect an attempt to verify the PIN.

A failed attempt to verify the PIN will be reported by the status word "Verification failed" (63 xx H, where xx indicates the number of tries remaining). If the PIN is blocked, the card will answer with "Authentication method (PIN) blocked" (69 83 H). Successful PIN verification is reported by the status word 90 00 H.

Here is a summary of the error status words that may be returned in response to the failed VERIFY command:

SW	Meaning
6A 86 H	Wrong parameters P1 and P2
67 00 H	Wrong length
69 83 H	Authentication method (PIN) blocked
63 xx H	Verification failed, xx tries remaining

If the PIN verification was successful, the card application must also update a field of the application sequence flag object, that is, set the field PIN Verified.

14.1.8 Application Cryptogram

After performing cardholder verification, the terminal proceeds to the terminal risk management process. During this procedure, the terminal analyzes the results of the previous transaction steps and comes to a decision about whether the transaction is to be approved, completed with the issuer online, or declined. For our sample application, the terminal risk management routine is quite simple and is described in detail in Chapter 16, where the terminal application is presented.

However, the terminal alone cannot make a decision about how the transaction is to be completed; its decision has to be approved by the card application and changed in some cases if the card so decides. Therefore, the terminal expresses its decision in the form of a request for a cryptogram that is sent to the card in the first GENERATE AC command. The terminal can request one of the following three cryptogram types:

- Transaction certificate (TC) for approving the transaction;

- Application authentication cryptogram (AAC) for declining the transaction;

- Authorization request cryptogram (ARQC) for completing the transaction online.

As already noted, our sample application will not support online services, so completing the transaction online is not possible. For this reason, the ARQC is left outside the scope of our sample application and the second GENERATE AC command, which is normally used to complete the transaction online, becomes unnecessary. Hence we remain with just two cryptogram types, TC and AAC.

If the card receives a request to decline a transaction, that is, a request for an AAC, it has no choice but to answer with AAC. However, if the card receives a request to approve the transaction, that is, a request for a TC, it has a right to answer with an AAC, that is, to decline the transaction if the results of its own card risk management demonstrate that the transaction cannot be approved.

Figure 14.3 presents a diagram of the GENERATE AC processing by our EMVdemo application. During the first step, we determine exactly what kind of cryptogram the terminal has requested. If the terminal has requested an AAC in order to decline the transaction, the card application does no special processing except setting proper fields in the card verification results (CVR) object. This object has quite an interesting purpose—it accumulates all results of transaction verification as well as the final decision of the card. The object is a part of the card's response to the GENERATE AC command and gives the issuer a chance to determine why the transaction has failed. The CVR object structure is explained in Section 14.2.1 in more detail.

If the terminal requests a TC in order to approve the transaction, the card application must perform the card risk management. The card risk

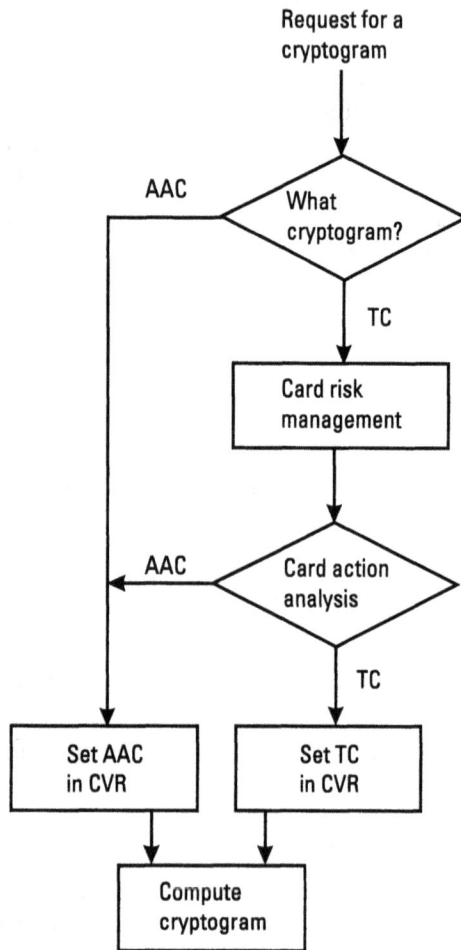

Figure 14.3 Diagram of GENERATE AC processing by EMVdemo.

management is used to analyze various application- and transaction-related parameters in respect of the transaction completing. Its results are accumulated in the CVR object.

In the next step of the request processing, card action analysis, the card application checks all results obtained during the card risk management process and verifies them against a set of preconditions for approving or declining the transaction. This set of preconditions, also referred to as card issuer action codes, is previously defined by the payment application issuer. The result of the card action analysis is the final card decision either to approve the transaction or to decline it. On the basis of its decision, the card updates relevant fields of the CVR objects and computes the application cryptogram authenticating the decision done by the card.

The card risk management routine, card action analysis, and the application cryptogram calculation algorithm are explained in Section 14.2 in detail.

We now return to the GENERATE AC command. It is sent in the form of an APDU having the following format:

CLA	INS	P1	P2	L$_c$	Data	L$_e$
80	AE	RCP	00	Transaction data length	Transaction data	00

Parameter P1 contains the RCP specifying a type of cryptogram requested. Bits 6–1 are reserved for future use and must all be set to 0. Bits 8–7 are coded according to Table 14.5.

The Data field of the command contains transaction-related data. Note that the transaction data are not coded as a TLV object. Therefore, the card must already know the order and length of particular items in the transaction data. Usually, the order and the length are defined on a payment system level and are therefore mandatory for all card applications and terminals operating within the system.

In our opinion, this approach significantly decreases the interoperability of payment applications as well as the interoperability between different payment systems. We can offer no explanation as to why this approach was taken.

For our sample application, we defined the transaction data field according to the Europay specification [5]. The transaction data for our

Table 14.5
Coding of the Leftmost Bits of the GENERATE AC RCP

Bit 8	Bit 7	Meaning
0	0	AAC requested
0	1	TC requested
1	0	ARQC requested
1	1	Reserved for future use

application is defined according to Table 14.6 in the order in which they appear.

A few remarks should be made on some data items included in the transaction data of Table 14.6. "Unpredictable number" is the random number generated by the terminal and sent to the card. This number is used as an input for a cryptographic method used to derive the session key for the application cryptogram calculation.

The terminal verification results consist of data on results of the terminal risk management and action analysis. Like any other field except "Amount authorized" and "Unpredictable number," the terminal verification results are not processed by our EMVdemo application and are used as an input to the cryptogram calculation algorithm.

Before proceeding to the card risk management and the cryptogram calculation, our card application will do some primary checks. First of all, if the parameter P2 does not match the predefined value, the card will answer with the status word "Wrong parameters P1 P2" (6A 86 H). Then, if the length of the Data field, that is, the length of the transaction data, is different from the expected value, the card will answer with "Wrong length" (67 00 H).

As a deviation from the EMV specifications, EMVdemo will report an attempt to request ARQC with the error status word "Wrong parameters P1 P2" (6A 86 H).

In the next step, the card application has to verify the fields of the application sequence flag object. If Get Processing is reset or AAC Generated is already set, the card will answer with the status word "Conditions of use

Table 14.6

Transaction Data

Data	Length
Amount, authorized	6
Amount, other	6
Terminal country	2
Terminal verification results	5
Transaction date	3
Transaction type	1
Unpredictable number	4
Terminal type	1
Data authentication code	2

are not satisfied" (69 85 H). Verification of the application sequence flag object fields will ensure consistency of the transaction flow.

After all primary verifications are successfully completed, the card application proceeds to the card risk management and the application cryptogram calculation. In response to the GENERATE AC command, the card will send not only the resulting cryptogram but also a rather significant set of other data objects. All of them are packed into one big TLV object whose content is defined in Table 14.7.

The cryptogram information data object basically contains the type of cryptogram computed by the card, that is, it reflects the card decision. Bits 8–7 are coded in the same manner as for the RCP. Other bits provide information on other events (PIN try limit exceeded, issuer authentication failed, etc.). For reasons of space, we omit a detailed description of other bit coding. An interested reader can find this information in various EMV specifications.

The application transaction counter object simply contains a value of the transaction counter maintained by the card application. The application cryptogram object contains the 8-byte cryptogram computed by the card.

The issuer application data object has a proprietary content and is defined by a particular payment application issuer. For our application, we follow the Europay specification [5] and define the content of this object as follows:

- Key derivation index (1 byte) and cryptogram version number (1 byte): parameters related to the cryptographic algorithms used to derive the session key and compute the cryptogram;
- Card verification (CVR) object (4 bytes);
- Data authentication code (DAC) object (2 bytes).

Table 14.7
Content of Data Object of the Response to the GENERATE AC Command

Object	Tag	Length
Cryptogram information data	9F 27 H	1
Application transaction counter	9F 36 H	2
Application cryptogram	9F 26 H	8
Issuer application data	9F 10 H	8

The DAC object is a static (i.e., precalculated) value authenticating the card application. It is computed during the card personalization stage and put on the card.

14.2 Security Functions

This section describes the security functions enforced by the card application in order to carry out a transaction. We start with a description of the card risk management and card action analysis procedures. Then, we explain how a session key for a cryptogram calculation is derived. The section concludes with a description of the cryptogram calculation algorithm.

14.2.1 Card Risk Management

Card risk management is a procedure whose goal is to define what type of cryptogram the card should send to the terminal. The overall card risk management procedure is formed by a group of independent card risk management functions. Each function of the groups concentrates on analyzing one particular aspect of the transaction flow. EMV specifications do not define a strict list of risk management functions that a card application must support. Instead, they recommend a list of functions that may be implemented in a particular payment system. A system developer may choose from this list some functions to implement in the system, or may also implement his own functions, keeping in mind interoperability issues.

Our EMVdemo application will support a limited number of card risk management functions, namely the following:

- PIN verification status;
- Maximum transaction amount;
- Cumulative transactions amount.

The PIN verification status function verifies whether the PIN verification was performed at all, whether it was successful, and whether the PIN is blocked. If the verification fails, the function reflects this fact by setting respective fields in the card verification results object.

For any payment application, a maximum has been set on the amount that can be paid within one single transaction. For our application, we define the maximum transaction amount to be 650 Euros. Verification of whether a transaction amount exceeds the predefined maximum is the task of the maximum transaction amount function. If the function detects that the transaction amount exceeds the maximum, it reflects this fact by setting respective fields in the card verification results object.

The last card risk management function that we will implement is the cumulative transactions amount function. This function ensures that the total amount of all transactions processed by a card does not exceed a maximum amount predefined by the card issuer. We define a maximum cumulative transaction amount for our sample application also to be 650 Euros. Analogously to other functions, the results of this function are reflected in the CVR object.

The CVR object comprises the results of the execution of all card risk management functions. The CVR object is composed of 4 bytes. The first byte contains the length of the object, which is obviously 04. Each individual bit of the other 3 bytes is responsible for reflecting a particular result of the card management routine. We use the coding of the last 3 three CVR bytes according to the Europay specification [5], which is shown in to Tables 14.8, 14.9, and 14.10.

Table 14.8
CVR Coding of Byte 2

b8	b7	b6	b5	b4	b3	b2	b1	Meaning
x	x							Type of cryptogram returned in the second GENERATE AC command: 00: AC 01: TC 10: Second GENERATE AC is not requested 11: Reserved for future use
		x	x					Type of cryptogram returned in the first GENERATE AC command: 00: AC 01: TC 10: ARQC 11: Reserved for future use
				1				Issuer authentication failed
					1			Offline PIN verification performed
						1		Offline PIN verification failed
							1	Unable to go online

Table 14.9
CVR Coding of Byte 3

b8	b7	b6	b5	b4	b3	b2	b1	Meaning
1								Last online transaction was not completed
	1							PIN try limit exceeded
		1						Exceeded velocity checking
			1					New card
				1				Issuer authentication failure on last transaction
					1			Issuer authentication not performed after online transaction
						1		Application blocked by card because the PIN try limit exceed the maximum
							1	Off-line static data authentication failed on last transaction

Table 14.10
CVR Coding of Byte 4

b8	b7	b6	b5	b4	b3	b2	b1	Meaning
x	x	x						Number of script commands processed successfully
			0					Not used
				1				Issuer script failed
					1			Lower consecutive offline limit or lower cumulative offline transaction amount exceeded
						1		Upper consecutive offline limit or upper cumulative offline transaction amount exceeded
							0	Not used

Because our application does not use all card risk management functions recommended by Europay, only the bits related to the functions used can be set.

After the card risk management is completed, it is time to perform the card action analysis. The card action analysis verifies the CVR object, and the result of this verification will be a decision on how the transaction is to be completed. In our case, the decision could be either to approve the transaction (to compute TC) or to decline the transaction (compute AAC).

The card action analysis procedure is based on another data structure: card issuer action codes (CIACs). CIACs are in fact bytes coded according to Tables 14.8 through 14.10 and expressing the condition when the transaction must be declined.[4] Their value is predefined by the payment system issuer. In other words, the CVR is simply compared against the CIACs and if at least 1 bit matches, the resulting decision will be to decline the transaction.

14.2.2 Session Key Derivation Algorithm

An application cryptogram is computed using DES, a symmetric cipher. Cryptographic keys used to compute the cryptogram are newly derived for each cryptogram and are referred to as session keys. Session keys are derived by a rather simple algorithm described in this section.

The session key derivation function is based on the Triple DES (DES3) cipher. For a session key derivation, a 16-byte DES3 key is needed. This key is computed and put on the card during card personalization. It is used only for deriving session keys and never for data encryption. Therefore, such a key is also referred to as a *master key* and denoted as MK.

A common input to compute a session key is an 8-byte random number

$$R = (R_0 \parallel R_1 \parallel R_2 \parallel R_3 \parallel R_4 \parallel R_5 \parallel R_6 \parallel R_7)$$

received from the terminal in the GENERATE AC command (R_i stands for each separate byte of the random number R).

The key derivation function computes two distinct session keys SK_l and SK_r (left and right) that are used in the cryptogram calculation. To produce two distinct keys using the same input and the same cipher, the third

---

4. EMV specifications distinguish between three types of action codes. Each type specifies conditions when the transaction must be approved, declined, or completed online, respectively. In our application, we use only the action codes of the type "decline."

most significant byte of the input random number is modified—for computing SK_l it is set to F0 H, for computing SK_r it is set to 0F H. Hence, a pair of session keys is computed using the modified random number and the master key MK as

$$SK_l = DES3(MK) \; [R_0 \parallel R_1 \parallel F0 \parallel R_3 \parallel R_4 \parallel R_5 \parallel R_6 \parallel R_7]$$

$$SK_r = DES3(MK) \; [R_0 \parallel R_1 \parallel 0F \parallel R_3 \parallel R_4 \parallel R_5 \parallel R_6 \parallel R_7].$$

Derived session keys are passed to the application cryptogram calculation algorithm.

14.2.3 AC Calculation Algorithm

In reality, the AC is a kind of a MAC that authenticates the card decision on completing a transaction. The input to the AC algorithm, that is, a message to authenticate, is formed by concatenation of the Transaction data received in the GENERATE AC command Data field and of a number of objects maintained by the card application. The complete list of input data for the AC calculation is given in Table 14.11.

Table 14.11
Input Data to the AC Calculation

Data Object	Supplied by
Amount authorized	Terminal
Amount other	Terminal
Terminal country code	Terminal
Terminal verification results	Terminal
Transaction currency code	Terminal
Transaction date	Terminal
Transaction type	Terminal
Unpredictable number	Terminal
Application interchange profile	Card
Application transaction counter	Card
Card verification results	Card

An input message M is a binary string composed of values of all objects listed in the table in the order in which they appear. The resulting input message for the algorithm must be a multiple of 8 bytes. Therefore, if the initial message is not a multiple of 8 bytes, it is padded according to ISO/IEC 9797 [6]—the last 8-byte block of the message is extended by byte 80 H on the right and then completed with a required number of bytes 00 H. The padded message is then split into 8-byte-long blocks labeled D_i.

The first step of the AC computing algorithm operates on the input message blocks of 8 bytes and is actually a block cipher encryption in the cipher block chaining (CBC) mode. On each iteration, the algorithm takes sequentially an input block D_i, encrypts it using DES and a key SK_1, and then XORs (exclusive-OR), the encryption result with the next block D_{i+1} (see Figure 14.4). The result of the XOR operation (i.e., addition modulo 2) is an input for the next iteration. This encryption procedure is done on all input message blocks except the last one.

The result of the last block encryption is decrypted using DES and another session key SK_r. The decryption result is in turn encrypted, again using the key SK_1. The result of the last encryption operation is the application cryptogram that had to be computed.

Figure 14.4 shows the AC calculation algorithms for a message consisting of five blocks of 8 bytes each, where "I" denotes an input to an operation; "O", an output; "DES e," an encryption operation; "DES d," a decryption operation; and ⊕, an XOR operation.

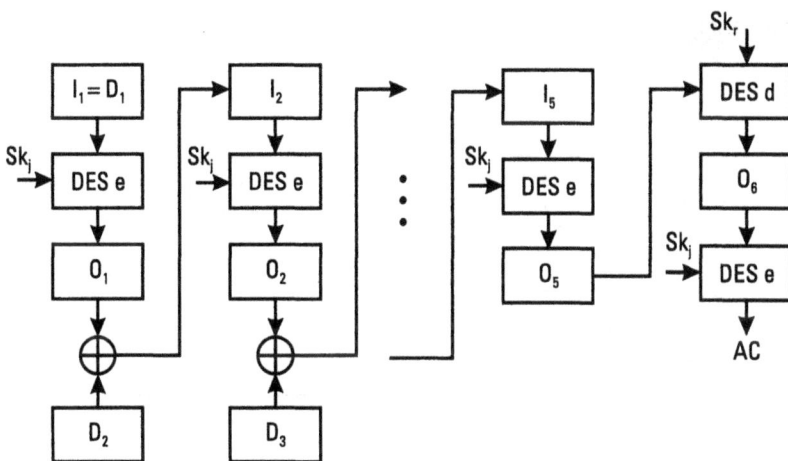

Figure 14.4 AC algorithm.

14.3   Application Design

Previous sections of the chapter were dedicated to a functional description of our sample application. Now is the time to present an architectural description of the application. The goal of this section is to give a brief introduction to the software modules of the application and to explain the distribution of overall application functionality among them. This section should form a logical link between the chapters describing the EMVdemo applet, the application OCF card service, and the terminal application.

Figure 14.5 presents a general overview of the sample application architecture. Obviously, all functionality related to the card application is inherited in the Java Card applet located on a Java Card technology smart card. The applet is run on top of the JCRE and is capable of performing all EMV card operations defined for our sample application. However, we will show in the next chapter that the applet architecture is not monolithic. The applet consists of several objects carrying out their own distinct tasks.

The terminal software is composed of two different layers. The OCF card service is located on the lower layer. It will carry out communication-specific functions and will provide the terminal with a uniform and communication-command-independent interface to EMV-related services provided by our card applet.

The terminal application is located on the upper layer. It inherits all EMV terminal functionality defined for our sample EMV credit/debit application. The terminal application will initiate and run a transaction and perform terminal-specific tasks like terminal risk management and terminal action analysis.

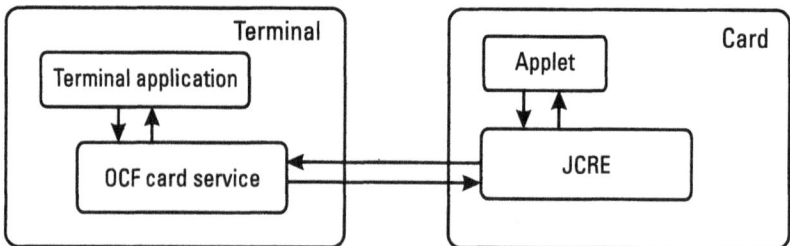

Figure 14.5 Sample application software architecture in general.

References

[1]   ISO 3166, *Codes for the Representation of Names of Countries,* Geneva: International Organization for Standardization, 1993.

[2]   ISO 4217, *Codes for the Representation of Currencies and Funds,* Geneva: International Organization for Standardization, 1990.

[3]   ISO/IEC 7813, *Identification Cards—Financial Transaction Cards,* Geneva: International Organization for Standardization, 1990.

[4]   "Minimum Card Requirements for Issuance of Chip Pay Now (Debit) and Pay Later (Credit) Cards," *Europay International,* Vol. 2.1, Oct. 1999.

[5]   Europay International S.A. "Off-the-Shelf Card Profile," Oct. 1999.

[6]   ISO/IEC 9797, *Information Technology—Security Techniques—Data Integrity Mechanism Using a Cryptographic Check Function Employing a Block Cipher Algorithm,* Geneva: International Organization for Standardization, 1993.

15

Java Card Applet Development

This chapter describes the Java Card applet implementing our sample EMV card application. It begins by demonstrating how the overall card application functionality is distributed among and represented on the basis of a number of Java classes. It then presents each class implementing the sample card application in detail.

15.1 Applet Architecture

The sample card application will have to carry out a number of different tasks. First it will have to perform the most basic tasks of any smart card application, namely, process incoming command APDUs and prepare and send out responses to the commands. Second, the card application will inherit functions related to the EMV card application. In addition, the card will have to manage various internal data objects representing particular aspects of the EMV card application.

Because the card applet will have to deal with different kinds of tasks, we will implement each particular type of task as a separate Java class. This will make the card applet highly modular and will increase its flexibility and extendibility. The last property is quite important because, as mentioned in Chapter 14, our sample EMV application does not implement all payment application functions defined in the EMV specifications. It would thus need to be extended for other applications in the future.

In accordance with the separation principle, we will implement the following tasks as separate Java classes:

- Java Card applet main class will handle all communication-specific tasks such as receiving command APDUs, processing them, and constructing and sending out responses.

- All EMV-specific functions will be represented in a separate class to be instantiated in the card applet object.

- Java Card 2.1 provides no support for an on-card file system and files (see Chapter 8). In contrast, an EMV card application should store many data objects in files. The most natural way to fill this need is to implement a file system ourselves in an object-oriented manner.

- In addition, the card risk management and card action analysis routines (see Chapter 14) will have to deal with a quite complex data object—the card verification results (CVR). It is therefore a good idea to implement CVR in the form of a separate class.

Figure 15.1 illustrates a general class architecture of our sample card application, a Java Card applet. The core of our card applet is a class covering all functions related to commands processing. We call this class EMVdemo. The EMVdemo class extends a standard Java Card class Applet, that is, its instance will be an ordinary Java Card applet executed by the on-card JCRE.

All EMV-specific functions will be implemented in another class, EMVPurse. The applet class EMVdemo will have to create an instance of this class in its constructor method to get access to data and functions of our

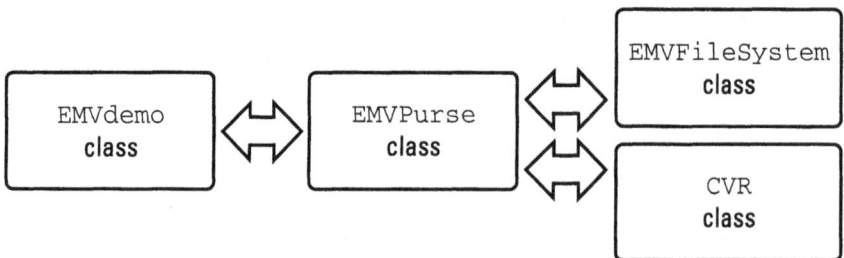

Figure 15.1 Card applet class architecture.

sample EMV card application. The interaction between the objects will be done through method invocation. This means that the EMVdemo object[1] will access functions provided by the EMVPurse object by invoking its respective methods.

The on-card file system will be implemented in the form of a separate class called EMVFileSystem. This class will provide an object-oriented implementation of a smart card file system. The class will provide essential means for working with smart card record files: creating files, writing records to a file, and reading records from a file. The record files are also to be implemented in the form of Java classes, where each record of a file will be represented by a separate class. The file system and the files are to be maintained by the EMVPurse object.

The CVR data object will also be implemented as a separate Java class called CVR. Apart from the data fields, the CVR class will provide a set of handy methods for modifying the fields in a data-structure-independent manner and reading the value of the fields. Because the card risk management and card action analysis are part of the EMV functionality, the CVR class is to be instantiated in the EMVPurse object.

All card application classes are grouped in the emvdemo package.

15.2 EMVPurse Class

The EMVPurse class implements all EMV-related functionality of our sample card application. This section describes in detail its internal data structures, the class constructor, the most important methods covering the major steps of the transaction flow, and other miscellaneous methods of the class. The class extends no other Java Card class, so its declaration will simply be

```
package emvdemo;
import javacard.framework.*;
import com.gieseckedevrient.javacardx.crypto.*;
public class EMVPurse {
...
}
```

Like any other class belonging to our sample card application, the EMVPurse class is assigned to the emvdemo package. We will have to

---

1. That is, an instance of the EMVdemo class.

import standard Java Card classes from the package `javacard.frame-work` and cryptographic support classes from the package of our smart card manufacturer[2]: `com.gieseckedevrient.javacardx.crypto`.

15.2.1 Data Structures and Related Methods

A detailed description of various EMV application–specific objects that must be managed by our card application was given in Chapter 14. Because all of those objects are strongly related to the EMV functionality of our card application, they should be defined and managed in the `EMVPurse` class. Our sample application does not cover the card personalization procedure. Therefore, we will assume that all static application data objects (see Section 14.1.3), that is, AIP, AFL, and the data objects referenced in AIP, are constants, and we will hard code their values in the class fields declaration section. For reasons of space, we do not demonstrate their declaration here; the interested reader can refer to Appendix A, where the complete code of the card application is presented.

Another constant object will require some explanation, however. The PIN value is also defined as a constant (`DEMO_PIN`) in the fields declaration section of the class. For simplicity's sake, we define the PIN value in a format in which it is expected to be received with the VERIFY command (see Section 14.1.7), that is, the constant representing it will contain not just the 4-digit PIN value but all 8 bytes, including the control field, the PIN length and also filler digits:

```
private final byte [] DEMO_PIN = {(byte) 0x24, (byte) 0x12,
                                  (byte) 0x34, (byte) 0xFF,
                                  (byte) 0xFF, (byte) 0xFF,
                                  (byte) 0xFF, (byte) 0xFF };
```

The master key value `MKac_VALUE`, used to derive session keys for the cryptogram calculation (see Section 14.2.2), is also defined as a constant in the fields declaration section of the class. Its declaration looks like this:

---

2. The company Giesecke & Devrient kindly provided the Sm@rtCafé Java Card toolkit that we used to develop our sample card application.

```
private final byte[] MKac_VALUE =
    { (byte)0x01, (byte)0x33, (byte)0x02, (byte)0x11,
      (byte)0x01, (byte)0x33, (byte)0x02, (byte)0x11,
      (byte)0x11, (byte)0x33, (byte)0x02, (byte)0x01,
      (byte)0x11, (byte)0x33, (byte)0x02, (byte)0x01 } ;
```

Note that we define all constants as private because they must be directly accessible only from the class to which they belong and none other, not even from another class located in the same package.

Apart from the constant values, there is a group of data objects that either has a complex character or whose value changes over the card application lifetime. In general, those data objects relate to either card security operations or transaction processing. Among the objects related to transaction processing, we can name the ATC and the cumulative transactions amount. They are defined as follows:

```
private short ATC = (short) 0;
private short cumulativeAmount = 0;
```

The ATC value must be accessible not only from the class EMVPurse methods; the methods of the EMVdemo class should also be able to read the ATC value and increment it by 1. For this purpose, we define two methods in the EMVPurse class that will let it perform those operations. The methods are defined as follows:

```
// Method getATC: returns the current value of ATC
public short getATC() {
    return ATC;
}
// Method incrementATC: increments ATC value by 1
public void incrementATC() {
    ++ATC;
}
```

We will represent the application sequence flag object as a binary array. Each byte element in the array will represent a separate field of the object. In the declaration sections, we define a reference to the array that will be created in the class constructor as a Java Card transient data structure:

```
public byte[] sequenceFlag = null;
```

For simplicity's sake, we define the application sequence flag as a public field of the class that can be accessed directly from any other card application class. This simplifies its handling.

The card application PIN will be represented and managed as a Java Card object instantiated from the standard Java Card `OwnerPIN` class. This class, as mentioned before (see Section 11.1.4), inherits all functionality related to PIN handling and provides a set of handy methods. In the class declaration section, we define a reference to our card application PIN object that is called `emvPIN` and will be used throughout the card application:

```
private OwnerPIN emvPIN;
```

We define `emvPIN` as a private field of the class because it must be managed only in the `EMVPurse` class and never accessed directly from any other class. However, some methods of the `EMVdemo` class would need access to the `emvPIN` object. We will therefore define a set of class methods that will allow access to the PIN object from "outside" the class. In fact, all of those methods are just wrappers around `OwnerPIN` class methods, that is, they simply invoke a respective method of the class and return the result. For instance, the wrapper method allowing verification of the PIN will look like this:

```
public boolean checkPIN(byte[] buffer, short offset,
                byte len) {
    return emvPIN.check(buffer, offset, len);
}
```

and the method returning the number of remaining PIN verification tries as follows:

```
public byte getPINTriesRemaining() {
    return emvPIN.getTriesRemaining();
}
```

To perform cryptographic operations in the application cryptogram calculation procedure, we would need a few more Java Card objects. The first objects are instances of the `SymmetricKey` object representing a cryptographic key to be used in symmetric cryptographic algorithms. The key objects are used in turn by cipher objects performing the desired cryptographic operations.

We will use two different ciphers in the application cryptogram calculation procedure: DES and Triple DES. Therefore, we have to define

references to DES and Triple DES key objects as well as the references to the
cipher object CipherECB:

```
private SymmetricKey desKey;
private SymmetricKey des3Key;
private CipherECB cipherDES;
private CipherECB cipherDES3;
```

The objects will be created and fed with initial values in the class constructor.

15.2.2 Class Constructor

Basically, two kinds of actions must be performed in the class constructor.
First, we have to create all objects that will be ever used by the applet during
its lifetime. Second, we must feed newly created objects with initial values if
required.

The first object is the application sequence flag. We will create it as the
transient byte array:

```
sequenceFlag =
    JCSystem.makeTransientByteArray(EMV.SEQF_LENGTH,
                    JCSystem.CLEAR_ON_RESET);
```

The transient array is created using a static method makeTransient-
ByteArray of the class JCSystem. The array length is defined by the constant SEQF_LENGTH located in the abstract interface EMV. The content of
the array is to be cleared each time that the card is reset. This is defined by
the CLEAR_ON_RESET constant. It will therefore be assured on the JCRE
level that the content of the application sequence flag is cleared on each reset
of the card.

The application PIN object is also to be created and initialized in the
class constructor:

```
emvPIN = new OwnerPIN(EMV.PIN_TRY_LIMIT, EMV.PIN_MAX_LEN);
emvPIN.update(DEMO_PIN, (short)0, (byte)8);
```

The first line of code will create an instance of the Java Card class OwnerPIN
with the following parameters:

- The maximum number of retries is specified by the PIN_TRY_
  LIMIT constant.

• The maximum length of the PIN value is specified by the PIN_
  MAX_LEN constant.

Both constants are defined in the abstract interface EMV, which contains definitions of all EMV-related constants used application-wide. The second line
of the code will set a PIN value in a newly created emvPIN object using its
method update. The PIN value of length 8 will be copied to the object
from the DEMO_PIN constant array.

Cryptographic key and cipher objects must be also created. The following lines of code will create two different symmetric key objects:

```
desKey =
    new SymmetricKey((short) 8, JCSystem.CLEAR_ON_RESET);
des3Key =
    new SymmetricKey((short) 16, JCSystem.CLEAR_ON_RESET);
```

The first object referenced by the desKey field of the class will represent an
8-byte symmetric key to be used by the DES cipher. The second object referenced by the des3Key field of the class will represent a 16-byte symmetric
key to be used by the Triple DES cipher. The values of the keys are to be
cleared on each reset of the card. This option is determined by the second
parameter of the key class constructor set to the value CLEAR_ON_RESET.

The cipher objects are not newly created with a constructor. Instead,
we use a static method getInstance of the Java Card class CipherECB.
This method returns a reference to a class instance:

```
cipherDES = CipherECB.getInstance(Cipher.ENGINE_DES);
cipherDES3 = CipherECB.getInstance(Cipher.ENGINE_3DES);
```

The only parameter of the getInstance method specifies a cipher algorithm to be used by the object. As can be seen above, two different references
are obtained, one to a cipher object implementing the DES algorithm and
one to an object implementing the Triple DES algorithm.

In the constructor, we also create the card application file system with
two elementary files. After the files are created, they are fed with constant
application data objects that are referenced in AFL and should be retrieved
using the READ RECORD command. For now, we will omit a detailed
description of how the file system and the files are created. This will be
explained in one of the following sections dedicated to the class
EMVFileSystem.

Apart from the data objects described in this section, we will also create a number of binary arrays in the EMVPurse class constructor. Such arrays have a miscellaneous character and will be used (and reused) in a number of the methods of the class, mainly as binary buffers for containing intermediate and final results of various application routines.

15.2.3 GENERATE AC Command Processing

The overall processing of the GENERATE AC command is performed by the processAC_I method of the EMVPurse class. This method receives a whole data field of the GENERATE AC command as an input. The result of the method is a completed answer[3] to the command. The method declaration looks like this:

```
public byte[] processAC_I(byte request, byte[] cdata,
                    byte offset, byte len) {
    ...
}
```

The parameter request specifies the type of application cryptogram (TC, AAC, or ARQC) requested in the command. The binary array reference cdata provides access to the binary data containing the data field of the command. The offset and the data length in the buffer are specified by the parameters offset and length, respectively. The method will be invoked directly from the EMVdemo applet's process method.

The method will sequentially perform a number of steps in order to process the GENERATE AC command and to prepare the response:

1. Reset the CVR object.
2. Perform the card risk management and card action analysis routines.
3. Fill data to the binary array containing a response to the command.
4. Derive session keys.
5. Compute AC.
6. Fill the result to the array.
7. Return the result.

---

3. Excluding the status word 90 00 H that is added to the response by JCRE.

The CVR object, `appCVR`, is reset simply by invoking its respective method:

```
appCVR.reset();
```

The `appCVR` object is also declared as a private field of the class. Its detailed description is presented later in this chapter.

The card risk management and card action analysis routines are grouped to one class method `riskManagement`, which is invoked on the second step. The method receives the requested type of cryptogram and the command data field as input. The result of the method execution is a decision regarding which type of cryptogram the card will use to answer the GENERATE AC command. The method is invoked as follows:

```
action = riskManagement(request, cdata, offset, len);
```

After the method execution, the `action` variable will contain the final decision. The method will be described in detail in the following subsection.

After it has been decided with which type of cryptogram the card will answer, we can start filling the answer to the command with data objects (see Chapter 14) such as cryptogram information data, ATC, and CVR. This operation mainly involves copying binary data to the array that will contain the answer. The reader can find the complete code of the method in Appendix A.

The next major task is to derive a pair of session keys for the application cryptogram calculation. However, some minor tasks must be accomplished before the key derivation procedure. First we must compose an input for the key derivation method. The input is composed of the ATC and a random number generated by the terminal, separated by 00 00. The variable `rand` will contain an input for the key derivation method. Its value is composed by the following few lines of code:

```
Util.setShort(rand,(short)0,ATC); // put ATC
rand[2] = (byte)0; rand[3] = (byte)0; // add 00 00
// copy the terminal UN from the command data
Util.arrayCopy(cdata, (short)(ISO7816.OFFSET_CDATA +
          EMV.OFFSET_TUN), rand, (short)4,(short)4);
```

We must also set the 16-byte master key value in the `des3Key` object that will be used by the cipher object in the key derivation method. This is done using the `setValue` method of the key object:

```
des3Key.setValue(MKac_VALUE,(short)0,(short)16);
```

Afterward, we bind the cipher object `cipherDES3` to the key object `des3Key` in this manner:

```
cipherDES3.setKey(des3Key);
```

The `setKey` method is used to specify the key object to be used by a cipher object to perform cryptographic operations.

Now, we can proceed with the session key derivation. As mentioned in Chapter 14, two session keys are required. They are derived by means of the same algorithm, with just 1 byte in the input being modified to derive a particular session key. We define two class methods that are capable of deriving the session keys—one method for each key. As parameters, both of the methods receive a reference to the cipher object `cipherDES3` and the input value `rand`. The resulting session key will be located in a binary array specified by a reference given as the third parameter. Both methods are invoked as follows:

```
derive_SK1(cipherDES3, rand, SK1);
derive_SKr(cipherDES3, rand, SKr);
```

The procedure performed by the key derivation methods is rather simple: The method performs a modification of the input value and then performs Triple DES encryption. As an example, we demonstrate an implementation of just one method computing the SK_l session key:

```
private void derive_SK1(CipherECB cipher, byte[] input,
                byte[] output) {
    input[2] = (byte)0xF0;
    cipher.encrypt(input, (short)0, (short)8, output,
            (short)0, CipherECB.PADDING_ISO00);
    return;
}
```

As soon as the session keys SK_l and SK_r are derived, we will be able to compute the application cryptogram. The calculation is performed using a separate method `compute_ac`. The method takes as input the message to be authenticated, a key pair SK_l, SK_r, and references to the key and cipher objects to be used by the method. The result of the method execution—the application cryptogram—will be placed in the binary array specified by a reference given to the method as the last parameter cryptogram. The method invocation will look like this:

```
compute_ac(message, EMV.MESSAGE_LEN, SKl, SKr, cipherDES,
           desKey, cryptogram);
```

A more detailed description of the method's internals is given in one of the following subsections. Little need be said about the input parameter message that references an answer to the GENERATE AC command to be authenticated by the cryptogram. Its structure was described in Chapter 14 in detail (see Section 14.1.8). It is composed of both the data received from the terminal and the card application data. The reader can find the underlying code in Appendix A.

After the application cryptogram is computed, its value along with the remaining data objects are filled into the array containing the answer to the GENERATE AC command. This is the last step that the processAC_I method performs.

Figure 15.2 illustrates the relationship between the processAC_I method and other methods of the EMVPurse class used to process the GENERATE AC command and to compute the application cryptogram.

15.2.4 CVR Object

The purpose and structure of the CVR object was explained in Chapter 14 (see Section 14.2.1). We now show how it is implemented. Because the object's structure is not simple, we will implement it as a separate Java class supplied with a set of methods to create its particular fields. The CVR implementation will be the class CVR. The class extends no other Java Card class and hence its declaration is simple:

```
class CVR {
...
}
```

The main content of the class is a 4-byte value containing particular CVR fields, which we will define as a private binary array field of the class CVR:

```
private byte[] bytes;
```

The bytes array is created in the class constructor method. In addition, we will set the value of the first byte to 03 according to the EMV specifications

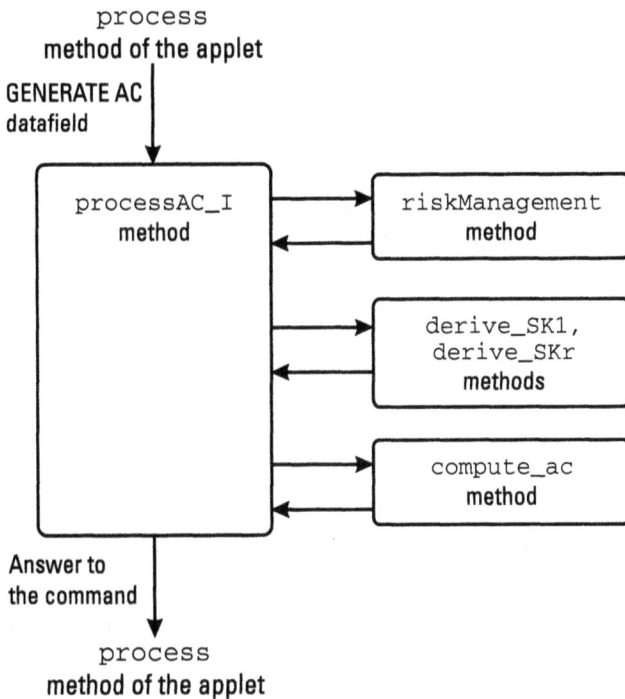

Figure 15.2 The processAC_I method and related methods diagram.

indicating the length of the object. The code of the CVR class constructor method is

```
public CVR () {
bytes = new byte[4];
    bytes[0] = 3;
}
```

What remains to be implemented is a set of methods that will read the CVR value and set a particular field. Reading the value is simple—we just have to return a reference to the internal bytes array:

```
public byte[] getBytes() {
    return bytes;
}
```

Reading a particular byte of the CVR object is just as simple:

```
public byte getByte (byte n) {
    return bytes[n];
}
```

We will also define a number of methods setting a particular field of the CVR object. For instance, in order to reflect in the CVR object that "Offline PIN verification was performed," we must set bit 3 in byte 2 (see Section 14.2.1). This action is implemented by the setPINperformed method, which is coded in the following way:

```
public void setPINPerformed() {
    // set bit 3 in byte 2
    bytes[1] = (byte) (bytes[1] | 4);
}
```

All other methods for setting a particular field in the CVR object are implemented in a similar manner. For reasons of space, we do not demonstrate their coding here. The reader can find the complete CVR class code implementation in Appendix A.

15.2.5 Card Risk Management and Card Action Analysis

Both the card risk management and the card action analysis routines of our sample card application are implemented in one single method, risk-Management, of the class EMVPurse. As an input, the method will receive a cryptogram type requested by the terminal and the whole data field of the GENERATE AC command. The result of the method execution will be the card decision on the cryptogram type to be returned to the terminal. Hence, the riskManagement method declaration will have the following form:

```
private byte riskManagement(byte request, byte[] cdata,
                            byte offset, byte len) {
    . . .
}
```

The first verification to be done by the method is to determine whether the terminal has requested the TC in order to approve the transaction or any other cryptogram type, that is, AAC or ARQC. If the terminal has requested the TC, the card proceeds with the card risk management and the card action analysis. Otherwise, no further verifications are done and the method can simply set the relevant CVR fields and return the answer, AAC. This simple verification routine is implemented in the following way:

```
// if ACC was requested, the card answers with AAC
// or if any other from TC was requested - answer with AAC
if ( (request == EMV.CODE_AAC) ||
                (request != EMV.CODE_TC) ) {
    // set relevant CVR bits
    appCVR.setGAC2notReq();
    appCVR.setAACinGAC1();
    return EMV.CODE_AAC;
}
```

If the terminal has requested the TC, we proceed with the card risk management functions. The first function performed by the method is the PIN status function. First of all, the function checks the status of the "PIN Verification Performed" field of the application sequence flag object. If the field is set, the respective field of CVR must also be set:

```
if (sequenceFlag[EMV.SEQF_PIN_PERFORMED] == TRUE)
            appCVR.setPINPerformed();
```

If the PIN verification fails, one more field of the CVR object must be set:

```
if ( (sequenceFlag[EMV.SEQF_PIN_VERIFIED] == FALSE) &&
(sequenceFlag[EMV.SEQF_PIN_PERFORMED] == TRUE))
appCVR.setPINFailed();
```

And, finally, if the PIN is already blocked, we must also reflect this application state in the CVR object:

```
if (emvPIN.getTriesRemaining() == 0)
    appCVR.setPINTryLimit();
```

The next card risk management function to perform is the "Maximum offline transaction amount," which checks whether the transaction amount exceeds a predefined value. In the first step of the function, we extract the transaction amount from the data received from the terminal. Then, if the transaction is in the native currency, we compare the transaction amount with the maximum allowed transaction amount. The function code is as follows:

```
// Maximum Offline transaction amount check function
        amount =
Util.makeShort(cdata[offset+4],cdata[offset+5]);
// if the amount exceeds the maximum
if ((Util.makeShort(cdata[offset+19], cdata[offset+20]) ==
     CURRENCY_CODE) && (amount  MAX_TRANS_AMOUNT))
appCVR.setMaxAmount();
```

If the transaction amount exceeds its maximum value, the respective field in the CVR object is set. We assume that the transaction amount is contained in the 4th and 5th bytes[4] of the data field, because it should not exceed an integer value 650. Nevertheless, we must still check that higher bytes are equal to 00:

```
if ( (cdata[offset] != 0) ||
     (cdata[offset+1] != 0) ||
     (cdata[offset+2] != 0) ||
     (cdata[offset+3] != 0) )
appCVR.setMaxAmount();
```

The last card risk management function, "Maximum cumulative amount," is similar to the previously described function. Its task is to ensure that the total cumulative amount of all transactions previously performed together with the current transaction amount will not exceed the predefined maximum value. This verification can be performed only if the requested transaction is in the native currency. The implementation of this function is shown below.

```
if ((Util.makeShort(cdata[offset+19], cdata[offset+20]) ==
     CURRENCY_CODE) &&
     (amount + cumulativeAmount  MAX_CUMUL_AMOUNT))
appCVR.setMaxAmount();
```

The cumulative amount of all transactions done with the card is stored in the cumulativeAmount private field of the EMVPurse class.

The last step is to perform the card action analysis procedure. This procedure is quite simple from a technical point of view: We just have to compare the CVR bytes with the card issuer action codes for declining the transaction (see Section 14.2.1). If at least one comparison matches, the transaction is to be declined; if not, the transaction is approved. The

---

4. Starting from 0.

following section of code demonstrates a complete implementation of the card action analysis procedure:

```
cvrBytes = appCVR.getBytes();
if ( ((cvrBytes[1] & CIACD[1]) == CIACD[1]) ||
// PIN Verification failed
((cvrBytes[2] & CIACD[2]) == CIACD[2]) ||
// PIN Try limit exceeded
((cvrBytes[3] & CIACD[3]) == CIACD[3]) ||
// Upper consecutive or
// Cumulative amount exceeded
(sequenceFlag[EMV.SEQF_PIN_PERFORMED] == FALSE))
// No PIN Verify
{
    // set relevant CVR bits
    appCVR.setGAC2notReq();
    appCVR.setAACinGAC1();
    action = EMV.CODE_AAC; // decline transaction
}
else
{
if (Util.makeShort(cdata[offset+19],cdata[offset+20])
== CURRENCY_CODE)
cumulativeAmount = (short)(cumulativeAmount + amount);
    // set relevant CVR bits
    appCVR.setGAC2notReq();
    appCVR.setTCinGAC1();
    action = EMV.CODE_TC; // approve transaction
                          // (issue TC)
}
```

If the transaction is approved, we must also increment the cumulative transactions amount with the amount of the current transaction. Whether the transaction is approved or not, the respective fields of the CVR object must be updated. The result of the card action analysis routine will be placed in the action variable, whose value will be returned as the result of the method execution.

15.2.6 Application Cryptogram Calculation

The AC calculation is performed by the compute_ac method. The method takes as input the message and its length mesLen, a key pair keyL, keyR, and references to the key and cipher objects, key and cipher. The result of the method execution—the application cryptogram—will be placed in the binary array specified by a reference ac given to the method as the last parameter. Thus, the method declaration will look like this:

```
private void compute_ac(byte[] message, byte mesLen,
                byte[] keyL, byte[] keyR,
                CipherECB cipher,
                SymmetricKey key,
                byte[] ac) {
    ...
}
```

Our AC calculation procedure strictly follows the specification presented in Chapter 14 (see Section 14.2.3). First we perform a calculation of the rounds number needed to calculate AC:

```
rounds = (byte)(mesLen / 8);
```

Then we initialize the key object with the first key of the key pair and bind the cipher object with the key object:

```
key.setValue(keyL, (short)0, (short)8);
cipher.setKey(key);
```

Then we copy the first 8 bytes of the message to the input register of the algorithm:

```
for (i=0; i 8; i++)
    ac[i]=message[i];
```

As an input register, we will use the ac array that will return the AC. We reuse that array in order not to create additional arrays and to minimize the memory usage.

Afterward, we are ready to start the iterative AC calculation that will be performed as many times as rounds were determined:

```
for (j=1; j (byte)(rounds+1); j++) {
    cipher.encrypt(ac,(short)(0), (short)8, ac, (short)0,
        CipherECB.PADDING_ISO00);
    if (j != rounds)
        for (i=0; i 8; i++)
            ac[i] = (byte)(ac[i]^message[(byte)(j*8+i)]);
}
```

In the last step, we perform decryption with the second key of the output data obtained after executing all rounds and then encryption with the first key:

```
key.setValue(keyR, (short)0, (short)8);
cipher.setKey(key);
cipher.decrypt(ac,(short)0, (short)8, ac, (short)0);

key.setValue(keyL, (short)0, (short)8);
cipher.setKey(key);
cipher.encrypt(ac,(short)(0), (short)8, ac, (short)0,
                         CipherECB.PADDING_ISO00);
```

At the end of the last encryption operation, the ac array will contain the calculated AC.

15.3 EMVdemo Class

We continue our card application description with the EMVdemo class. The class extends the Java Card class Applet and inherits from it all methods needed to register the applet with JCRE, process incoming APDUs, and handle the applet selection and deselection (see Chapter 8). Therefore, the class declaration will look like this:

```
package emvdemo;
import javacard.framework.*;
public class EMVdemo extends Applet {
...
}
```

Our EMVdemo class will have to implement the install, select, and process methods of class Applet. Descriptions of those methods are given in the following subsections.

The class will have just one field containing an instance of the class EMVPurse that implements all EMV-related functions of our sample card application. This field is declared as

```
private EMVPurse emvPurse;
```

The field is called emvPurse and is declared to be private because it must not be accessible for any class except the owner class, EMVdemo.

15.3.1 Class Constructor, Methods `install` and `select`

The first method that our applet has to implement is the `install` method, which is invoked by JCRE at the end of the applet installation procedure in order to register the applet with JCRE. This method creates an instance of the applet class EMVdemo. Hence, the method will look like

```
public static void install (byte[] buffer, short offset,
                               byte length)
{
    new EMVdemo();
}
```

The class constructor is as simple as the method `install`. In the constructor, we will have to create an instance of the EMVPurse class and perform a mandatory operation for any Java Card applet—registering the applet instance with JCRE. The code of the applet is rather short:

```
private EMVdemo()
    {
        emvPurse = new EMVPurse();
        register();
    }
```

We have no special actions to perform when the applet becomes selected. Therefore we will leave the implementation of the applet `select` method empty, ensuring that the method will always return `true` in order to let the applet be always selectable:

```
public boolean select (APDU apdu)
{
    return true;
}
```

The core of the applet implementation is the `process` method presented in the next subsection.

15.3.2 Method `process`

The `process` method of our applet is invoked by the JCRE each time an incoming command APDU is received by the card. The task of the method is to process the incoming APDU and to generate a response to it. The method has just one input parameter: the reference to the APDU object containing the received APDU. The declaration of the method is as follows:

```
public void process(APDU apdu) throws ISOException
{
    ...
}
```

The method can throw an ISOException indicating that a certain error occurred during the APDU processing, and the card must answer with the error status word specified by the exception.

First, we must obtain a reference to the binary array buffer containing the APDU byte using the getBuffer() method of the APDU class:

```
byte[] buffer = apdu.getBuffer();
```

It is good practice to check whether the received APDU just selects our applet or already contains a command dedicated to our card application. The method selectingApplet() of the class Applet can help us to do this. The method will return true if the received APDU is the SELECT APDU selecting our applet or false if any other command APDU was received. The only thing to do if our applet is currently being selected is to reset all fields of the application sequence flag object except the Application Selected field that must be set. The following fragment of code demonstrates an implementation of this action:

```
if (selectingApplet())
{
// Modifying the flags
emvPurse.sequenceFlag[EMV.SEQF_APP_SELECTED] = TRUE;
emvPurse.sequenceFlag[EMV.SEQF_GETPROC_PERFORMED] = FALSE;
emvPurse.sequenceFlag[EMV.SEQF_ARQC_GENERATED] = FALSE;
emvPurse.sequenceFlag[EMV.SEQF_AAC_GENERATED] = FALSE;
emvPurse.sequenceFlag[EMV.SEQF_PIN_PERFORMED] = FALSE;
emvPurse.sequenceFlag[EMV.SEQF_PIN_VERIFIED] = FALSE;
return;
}
```

The rest of the applet process method is organized in a form of nested Java switch statements. The first switch statement is used to distinguished APDUs with different CLA bytes. For each CLA case, a separate switch statement is used to make it possible to distinguish APDUs with the same CLA byte but with different INS bytes.

Our applet must support APDUs with two different CLA bytes—00 H defined according to ISO/IEC 7816 and 80 H defined according to the EMV specifications. To explain how the process method of our applet works, we will first show a "skeleton" of the nested switch statements and then explain each case in detail:

```
switch (buffer[ISO7816.OFFSET_CLA])
{
    case ISO7816.CLA_ISO:
        switch ( buffer[ISO7816.OFFSET_INS])
            {
                case EMV.INS_READ_RECORD:
                ...
                break;
                case EMV.INS_VERIFY:
                ...
                break;

                default:
                    ISOException.throwIt(
                        SO7816.SW_INS_NOT_SUPPORTED);
            }
            break;

    case EMV.CLA_MANUFACTURER:
        switch (buffer[ISO7816.OFFSET_INS])
        {
            case EMV.INS_GET_PROCESSING_OPTIONS:
            ...
            break;

            case EMV.INS_GENERATE_AC:
            ...
            break;

            default:
                ISOException.throwIt(
                    SO7816.SW_INS_NOT_SUPPORTED);
        }
        break;
```

```
default:
        ISOException.throwIt(
                        SO7816.SW_CLA_NOT_SUPPORTED);
}
```

The nested `switch` structure shown above will support all four kinds of command APDUs that must be processed by our card applet. Two commands have the CLA byte in accordance with ISO/IEC 7816; the other commands have the CLA byte in accordance with EMV specifications. If the applet receives a command APDU with the CLA or INS bytes that are not supported, it will answer according to ISO/IEC 7816 with the error status words "CLA no supported" or "INS not supported," respectively. The answer with the error status word will be generated using the `ISO-Exception`, which will be thrown in the `default` cases of all switch statements.

Figure 15.3 illustrates the process of distinguishing APDUs with different CLA and INS bytes and also gives an overview of the nested `switch` structure of the applet's `process` method. The next few subsections describe the processing of each command APDU. The order of the descriptions follows the main transaction steps.

15.3.2.1 GET PROCESSING OPTIONS Command

A transaction is started with the GET PROCESSING OPTIONS command APDU sent by the terminal to the card. The first thing that the applet should do is verify the command parameters and check the state of the application sequence flag object fields. The following fragment of code demonstrates the verification procedure performed by the applet:

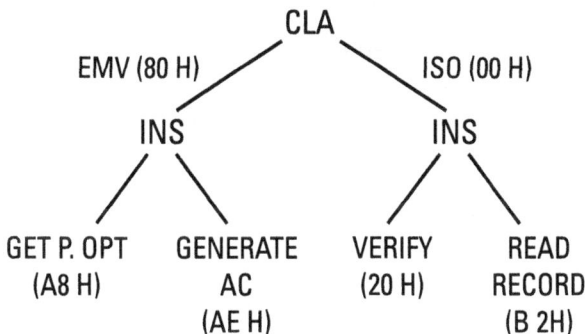

Figure 15.3 APDU processing diagram.

```
case EMV.INS_GET_PROCESSING_OPTIONS:

if ( (buffer[ISO7816.OFFSET_P1] != EMV.P1_GET_PROC_OPT) ||
     (buffer[ISO7816.OFFSET_P2] != EMV.P2_GET_PROC_OPT) )
          ISOException.throwIt(EMV.SW_WRONG_P1P2);

if ( buffer[ISO7816.OFFSET_LC] != EMV.LC_GET_PROC_OPT )
     ISOException.throwIt(ISO7816.SW_WRONG_LENGTH);

if (emvPurse.sequenceFlag[EMV.SEQF_APP_SELECTED] != TRUE ||
emvPurse.sequenceFlag[EMV.SEQF_GETPROC_PERFORMED] == TRUE )
     ISOException.throwIt(EMV.SW_COND_NOTSAT);

if ( emvPurse.getATC() == EMV.ATC_MAX_VALUE )
     ISOException.throwIt(EMV.SW_REFD_INVALID);
...
```

In addition, the ATC value must be checked. If it has already reached its maximum allowed value, the card application is invalidated and a respective error status word must be sent in the answer to the terminal.

If all verifications are successful, we increment the ATC value by one and change the status of certain fields of the application sequence flag:

```
// Incrementing the transaction counter
emvPurse.incrementATC();

// Setting the sequence flags
emvPurse.sequenceFlag[EMV.SEQF_GETPROC_PERFORMED] = TRUE;
emvPurse.sequenceFlag[EMV.SEQF_ARQC_GENERATED] = FALSE;
emvPurse.sequenceFlag[EMV.SEQF_AAC_GENERATED] = FALSE;
```

In the next step of the command processing, we fill the AIP and AFL data into the APDU object buffer referenced by buffer. Finally we send the response APDU with the data that we have placed into the buffer. This is done using the following method of the APDU class:

```
apdu.setOutgoingAndSend((short)0, (short)16);
break; // end of the case
```

The method will send the response APDU containing 16 bytes from the buffer starting at offset 00. JCRE will add the success status word 90 00 H itself.

15.3.2.2 READ RECORD Command

The READ RECORD command is used to read the content of the application elementary files referenced in AFL. As with any other command, we must first perform a verification of the command parameters:

```
case EMV.INS_READ_RECORD:
    if ( buffer[ISO7816.OFFSET_P1] == 0 )
        ISOException.throwIt(EMV.SW_FUNC_NSUPP);
    // last three bits must be 100 B
    if ( (buffer[ISO7816.OFFSET_P2] & 7) != 4)
        ISOException.throwIt(EMV.SW_FUNC_NSUPP);
    ...
```

Then we extract the file SFI and the record number from the command parameters P1 and P2:

```
sfi = (byte) (buffer[ISO7816.OFFSET_P2]>>>3);
rn = buffer[ISO7816.OFFSET_P1];
```

Afterward, we simply read the given record from the file specified by its SFI and determine the length of the record data:

```
record = emvPurse.readRecord(sfi,rn);
rl = record.getActualLen();
```

Note that we do no SFI and record number verification. If either a nonexistent file or a nonexistent record in the file is referenced, a respective ISO-Exception will be thrown by the file system or the file objects (their description will be given in the next section of the chapter) and the command processing will be terminated at that point.

If the record reading is successful, we merely have to fill the record data into the APDU buffer and then issue the response APDU:

```
Util.arrayCopy(record.getData(), (short)0,
                buffer, (short)0, (short) rl);
apdu.setOutgoingAndSend((short)0, (short)rl);
break; // end of the case
```

15.3.2.3 VERIFY Command

The VERIFY command is used to perform cardholder verification on the basis of an offline plaintext PIN verification. As with other command

processing, we must verify the command parameters P1 and P2. (We omit
the code because it is like that for the previously explained commands.) If the
verification is successful, we must update the content of particular fields of
the application sequence flag:

```
emvPurse.sequenceFlag[EMV.SEQF_PIN_PERFORMED] = TRUE;
emvPurse.sequenceFlag[EMV.SEQF_PIN_VERIFIED] = FALSE;
```

We must also verify whether the PIN is blocked:

```
if (emvPurse.getPINTriesRemaining() == 0)
        ISOException.throwIt(EMV.SW_AUTHM_BLK);
```

If all verifications are successful, we have to load command data into the
APDU buffer. This is done using the special method of the APDU object:

```
apdu.setIncomingAndReceive();
```

The APDU buffer will contain the command data after this method execution. (Initially, the buffer contains only the first 5 bytes of the command
APDU.)

Now we are ready to perform the PIN verification. The PIN data are
contained in the APDU buffer and can be referenced directly:

```
if (
emvPurse.checkPIN(buffer,(short)ISO7816.OFFSET_CDATA,
                      EMV.PIN_MAX_LEN) )
{
    emvPurse.sequenceFlag[EMV.SEQF_PIN_VERIFIED] = TRUE;
    return;
}
else
ISOException.throwIt(Util.makeShort(EMV.SW1_VER_FAILED,
                    emvPurse.getPINTriesRemaining()));
break; // end of the case
```

If the PIN verification was successful, we will update the application
sequence flag field. In the opposite case, the applet will return the error status
word "PIN verification failed." The second byte of the status word will contain the number of remaining tries to verify the PIN.

15.3.2.4 GENERATE AC Command

GENERATE AC is the last command that must be processed by the applet on the way to completing the transaction. Apart from loading the command data and command parameters P1 and P2 verifications, we must check the status of the application sequence flag in order to determine whether the card application is in the right state to complete the transaction:

```
// Verify the transaction context
if ((emvPurse.sequenceFlag[EMV.SEQF_GETPROC_PERFORMED]
        == FALSE) ||
(emvPurse.sequenceFlag[EMV.SEQF_AAC_GENERATED] == TRUE) )
// ... then return "conditions of use are not satisfied"
    ISOException.throwIt(EMV.SW_COND_NOTSAT);
```

We also check that it is not the second GENERATE AC command and that it is not the ARQC request (the respective code is omitted here).

In the next step we invoke the command processing method of the object emvPurse:

```
response = emvPurse.processAC_I(buffer[ISO7816.OFFSET_P1],
        buffer, ISO7816.OFFSET_CDATA,
        buffer[ISO7816.OFFSET_LC]);
```

We pass the reference to the APDU data buffer containing the command data and specify the data offset and data length. The result of the command execution, that is, the completed answer to the command including the AC, will be referenced by response. The remaining thing to do is simply to copy the answer to the command referenced by response to the APDU buffer and to send out the response APDU:

```
Util.arrayCopy(response, (short)0, buffer, (short)0,
            (short)(response[1]+2));
// reset sequence flags
emvPurse.sequenceFlag[EMV.SEQF_GETPROC_PERFORMED] = FALSE;
// send the response out
apdu.setOutgoingAndSend((short)0,
(short)(response[1]+2));
break; // end of the case
```

We merely have to reset the field of the application sequence flag object before sending the answer. The object field will indicate that the next transaction processing can be started.

15.4 EMVFileSystem **Class**

Current Java Card 2.1 specifications contain no classes for on-card file system support. For this reason we have had to make our own implementation of the card file system for our card application. The functionality provided by our implementation is narrowed to the specifics of our application—we will support only record files and will provide mechanisms for deleting existing files and records in the files.

Our file system implementation is object oriented, which means that each particular element starting from a record in a file is implemented in the form of a Java class. Figure 15.4 illustrates our file system implementation. The topmost level of the file system is the instance of the EMVFileSystem class. Its tasks are to provide a set of basic functions for working with the files. Each file maintained by the file system is an instance of the EMVFile class. In turn, records of any file are instances of the EMVFileRecord class. This section of the chapter provides a basic overview of our card file system implementation and avoids programming details.

15.4.1 Record

Our description begins with the lowest level of the file system—a file record represented by the EMVFileRecord class. The content of the record is an ordinary byte array that is defined as a private field of the class.

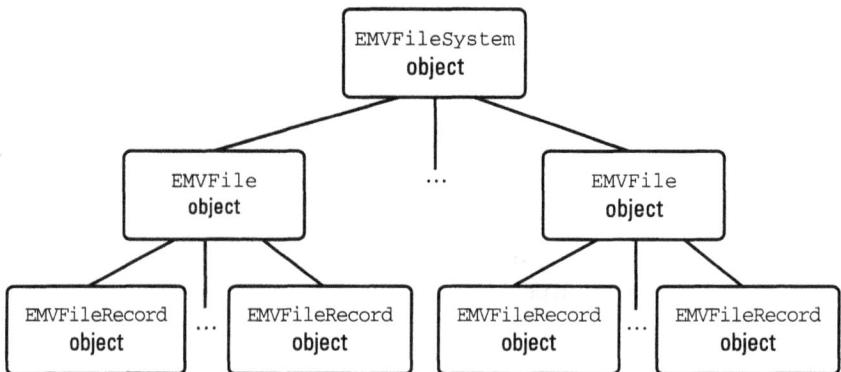

Figure 15.4 File system implementation.

```
class EMVFileRecord
{
private byte rec_length; // maximum length of the record
private byte act_length; // actual length of the record
private byte[] record_data;
...
}
```

The array is created in the class constructor. The array length is specified by the constructor parameter:

```
public EMVFileRecord (byte rl)
{
    record_data = new byte[rl];
    rec_length = rl;
}
```

The data are put into the record using the writeData method of the class having the following implementation:

```
public void writeData (byte[] value, byte len)
{
    JCSystem.beginTransaction();
    Util.arrayCopy(value, (short)0, record_data, (short)0,
                   (short)len);
    act_length = len;
    JCSystem.commitTransaction();
}
```

The data to be written to the record and the data length are received as the method parameters. The length of the data to be written is checked on a higher level, that is, in the file class. An interesting aspect of the method implementation is the transaction atomicity. Both operations—copying the data and setting the length— must be performed atomically, that is, either both operations must be completed or neither of them. We ensure atomicity using the means provided by the Java Card: The beginTransaction method will start the transaction and the commitTransaction method will complete it. The atomicity of the actions between the invocation of those two methods is guaranteed by Java Card.

Other methods of the class, like reading the record data, are so simple that we omit their code here.

15.4.2 File

The record file functionality is represented by the EMVFile class. The class maintains its records as an array of EMVFileRecord objects.

```
class EMVFile
{
    private byte rec_number; // number of records
    private EMVFileRecord[] file_records;
    private byte _SFI;
...
}
```

The record objects are created in the class constructor method. The constructor receives the number of records, the file SFI, and the record number as parameters. The whole code of the constructor method is given below:

```
public EMVFile (byte sfi, byte rn, byte rl) throws
ISOException
    {
        byte i;
        if (rn  EMV.MAX_FILE_SIZE)
            ISOException.throwIt(EMV.SW_MEM_FAILURE);
        _SFI = sfi;

        file_records = new EMVFileRecord[rn];
        for (i=0; i rn; i++)
            file_records[i] = new EMVFileRecord(rl);
        rec_number = rn;
    }
```

The constructor will throw the ISOException reporting the error "Memory failure" if the requested number of records exceeds a predefined value. This measure will help to prevent a situation where a card application creates a file so large that it occupies too much card memory.

Data are written to a file record using the writeRecord class method. The method checks both the record number and the length of the data to be written to the file. Any error will be reported by an ISOException with a respective status word. The fragment of code below demonstrates the method implementation:

```
public void writeRecord (byte recnum, byte[] value,
                         byte len)
```

```
            throws ISOException
{

        if (recnum > rec_number - 1)
    ISOException.throwIt(ISO7816.SW_RECORD_NOT_FOUND);
        if (len > file_records[recnum].getRecordLen() - 1)
    ISOException.throwIt(ISO7816.SW_WRONG_LENGTH);
        file_records[recnum].writeData(value, len);
}
```

If both the length and the record number are acceptable, the method will simply invoke a writeData method of the record object in order to write the data to that record.

The reading of record data is performed using the method read-Record, which is implemented in a similar way except that no length check is performed. This method, like other miscellaneous methods, is rather simple, so we will not go into its code here.

15.4.3 File System

The EMVFileSystem class is the topmost level of our file system implementation. The files are maintained by the file system as an array of the EMVFile objects. The array is defined as a private field of the class. The reference to the file objects is created in the class constructor method. The file objects themselves, however, are created only after a request to create a new file. The one parameter of the method is the maximum number of files that should be supported by the file system. The following fragment of code demonstrates the class declaration and the class constructor method implementation:

```
class EMVFileSystem
{
    private EMVFile[] files;
    private byte files_num; // Maximum files supported
    private byte next_av = 0; // next file to create

    public EMVFileSystem (byte maxfiles)
    {
        files_num = maxfiles;
        files = new EMVFile[maxfiles];
    }
    ...
}
```

We would like to point out that the complete implementation of our file system also supports selecting particular files. In this chapter, we provide no description of related data and mechanisms. However, an interested reader can find the code of the complete implementation in Appendix A of this book.

Files are created using the `createFile` method of the class. The method receives the SFI of the file to be created, the number of records, and the record length as parameters. The method simply verifies whether a new file can be created (that is, whether a maximum allowed number of files has been reached). If the file can be created, the method invokes a constructor of the file class. The code of the method is demonstrated below:

```
public void createFile (byte sfi, byte recnum, byte
reclen) throws ISOException
{
    if (next_av == files_num)
     ISOException.throwIt(ISO7816.SW_FILE_INVALID);
    files[next_av] = new EMVFile(sfi, recnum, reclen);
    next_av++;
    }
```

The file system implementation should also provide methods for reading data from the files and writing data to the files. Those functions are implemented by the `readRecord` and `writeRecord` methods. Because both methods are similar from the point of view of implementation, we will explain just one—the `readRecord` method. The method receives the SFI of the file and the record number as parameters. The method will return a reference to the complete record object located in the file specified by the given SFI and record number. If no file with the given SFI is found, the method will throw the `ISOException` with the status word "File not found." The method implementation is listed below:

```
public EMVFileRecord readRecord (byte sfi, byte recnum)
throws ISOException
   {
      byte i;
      boolean f = false;
      for (i=0; i<next_av; i++)
        if (files[i].getSFI() == sfi)
        {
           f = true;
           break;
        }
```

```
    if (!f)
      ISOException.throwIt(ISO7816.SW_FILE_NOT_FOUND);

    return files[i].readRecord((byte)(recnum - 1));
  }
```

If the file with the given SFI is found, the method will simply invoke the readRecord method of the object related to that file.

Our file system implementation presented in this section is rather compact. However, it demonstrates certain practical features of Java Card that make it possible to implement an object-oriented on-card file system.

16

OCF Card Service Development

This chapter shows how to build card services for smart card applications using OCF classes, and it should be implemented by someone who has detailed knowledge of smart card commands. A card service will usually be issued by a provider of a card applet in conjunction with the applet. It should have an easy-to-understand API in which application programmers can implement their own graphical user interfaces (GUIs). This chapter also covers deeper issues like card terminal factories.

16.1 Setting Up the Environment

First of all we need the OCF packages, which can be obtained from http://www.opencard.org/index-downloads.shtml. The all-in-one package for Version 1.2 should fit everyone's needs because it comes with documentation and sources. Two files must be available to our CLASSPATH-Variable, `base-core.jar` and `base-opt.jar`. The package `base-core.jar` contains all important classes necessary to develop the card's service, such as `SmartCard`, `CardTerminalRegistry`, and `CardServiceRegistry`. The package `base-opt.jar` contains additional classes required, such as `CardServiceUnexpectedResponseException` (which is thrown when a status word other than 90 00 H is received) or classes for handling TLV objects. An `opencard.properties` file must also be created, in which the factories that have to be used are specified.

Properties files can reside in several places in a UNIX environment (this would be some user's home directory); in a Windows environment, it is usually `%JAVA_HOME%\jre\lib`.

We then require classes for the card reader. They come with the development package. (Classes for a few readers, e.g., IBM or Gemplus, come with the OCF package, but are out of date. Because we have chosen a G&D reader, we cannot use them in any case. You should retrieve the latest drivers from your card terminal manufacturer.) The classes are needed for implementation of the `CardTerminal`, the `CardTerminalFactory`, and the underlying driver classes. The personal chipcard terminal (PCT) driver also has to be installed. This is part of the setup program provided with the toolkit. Normally a developer must simply include the `jar` file in the `CLASSPATH` and set the `CardTerminalFactory` property, but if you want to use the T = 1 protocol, a PTS has to be made and the correct transport manager invoked. Because this is vendor specific, we do not go into further detail here.

G&D drivers depend on the Java Communications API for communicating with the serial port, which can be obtained from the Javasoft home page.[1] The `comm.jar` file must be included in the `CLASSPATH`; the `javax.comm.properties` file and in some cases a low-level (platform-dependent) library also must be copied to the right location. For Windows operating systems, the file `win32com.dll` should reside in `%JAVA_HOME%\jre\bin`, for example. However, other drivers from other vendors can use another solution to communicate with the card terminal.

16.2   The Properties File and the Factory

In the `opencard.properties` file all properties regarding OCF are defined. First we must define at least one card terminal factory and one card service factory (more than one can be named, of course). The card terminal factory comes with the vendor package, but a card service developer will have to create her own suitable card service factory. The factory can identify that it can produce a corresponding card service object. For example, an `opencard.properties` file will look something like this:

---

1. http://www.javasoft.com/products/javacomm/index.html.

```
#####################################
# Card terminal/service configuration#
#####################################
OpenCard.terminals =
emvdemo.CardTerminal.GDT1CardTerminalFactory|EMVDemo|PC
T_T1_200|1
OpenCard.services =
emvdemo.CardService.EMVDemoCardServiceFactory
#############################
# Trace configuration        #
#############################
GieseckeDevrient.OpenCard.SerialIO.logLevel = 1
GieseckeDevrient.OpenCard.SerialIO.logFile = CardFile.log

OpenCard.trace = opencard:5 com:5 emvdemo:5
```

The first property defines the card terminal factory to use. The parameter fields, separated by the "|" symbol, define the following:

- The name of the class file;

- An arbitrary name for identifying the card reader for references;

- A name that the factory needs for deciding which object to create;

- A number that identifies the serial port (e.g., 0 means COM1, 1 means COM2).

The next property defines the card service factory. No parameter fields need to be provided. Next, some of the G&D-specific properties control the low-level debug output of the driver, that is, they control exactly what will be sent to the reader. For example,

```
[COM2]<- 6F 11 05 32 02 00 0D 00 A4 04 00 07 45 4D 56 64
65 6D 6F 00 F5
[COM2]-> 20 00 02 90 00 B2
```

is the output of a SELECT command with all the APDU bytes and card reader control bytes. The last line defines the trace level of an open-card.core.util.tracer object, which is created inside various Open-Card classes. Different levels are available in different packages, so the levels in the opencard.* packages can be different from those in com.* packages. There are eight levels of trace, from emergency (0) to lowest (8). If the level of the message is less than or equal to the given level, then tracing output is produced. For a complete list of the trace levels, see [1].

The main object in conjunction with the properties file is the `Card-ServiceFactory`. There is no need to implement the factory entirely since a generic (abstract) `CardServiceFactory` class is provided with OCF. Thus we can simply derive a class and implement the two abstract functions. A change has been made in moving from the OCF 1.1 to the OCF 1.2 implementation. The older version requires implementation of a `knows` method and a `cardServiceClasses` method. There is an `OCF11Card-ServiceFactory` in the `opencard.opt.service` package for backward compatibility. OCF 1.2 card service factories need a `getCardType` method and a `getClasses` method. The main advantage in OCF 1.2 is that `knows` can only put out a true/false decision if the factory knows the service and `getCardType` can deliver more specific information about the card type. Since Version 1.23, according to the version history, the method can communicate with the card via a `CardServiceChannel` object. This is important to multiapplication cards, since knows has only an ATR as a parameter, which is not sufficient in multiapplication cards. (Note that it is not always possible to change the ATR.) Thus the factory can determine if a specific application is on the card (for example, by selecting the AID) and that it is responsible for obtaining a card service for it. The card service factory suitable for our card service might look like this[2]:

```
package emvdemo.CardService;

import opencard.core.service.*;
import opencard.core.terminal.*;
import java.util.Enumeration;
import java.util.Vector;

public class EMVDemoCardServiceFactory extends
CardServiceFactory
{
    private static Vector services=new Vector();
```

This vector contains the classes the factory can provide. It has to be defined at initialization time or be declared a static variable.

---

2. Many of the initialization values are constants. The constants will not be explicitly mentioned in the text.

```
static
{
services.addElement(emvdemo.CardService.EMVDemoCard
Service.class);
}
```

Now we turn to the `getCardType` method. We make a SELECT-APDU and send it via the card scheduler to the card. If the application can be selected (and assuming there have been no formal mistakes such as setting the wrong parameters), the card is able to produce the concrete card service object.

```
    public CardType getCardType(CardID cid,
CardServiceScheduler scheduler)
    throws CardTerminalException
    {
    CommandAPDU Application=new
CommandAPDU(TransactionConstants.SC_CMD_HDR_LEN+
TransactionConstants.AID_LEN+1);
```

The length of the APDU is determined by the header length (which is the same for all smart card commands), the length of the data, and the length of the L_e byte. (SELECT does have return data but we discard it.)

```
Application.append(TransactionConstants.CLA_ISO);
Application.append(TransactionConstants.INS_SELECT);
Application.append(TransactionConstants.P1_SELECT);
Application.append(TransactionConstants.P2_SELECT);
Application.append(TransactionConstants.AID_LEN);
Application.append(TransactionConstants.TRANSACTION_AID
);
    Application.append(TransactionConstants.LE_SELECT);

    ResponseAPDU
res=scheduler.getSlotChannel().sendAPDU(Application);

    if (res.sw()==TransactionConstants.SW_OK)
    {
        return new CardType(1);
    }
```

Because there are no extra attributes to be added to `CardType` and we only support one card service, the answer is simply 1.

```
else
{
     return (CardType.UNSUPPORTED);
}
```

If the application is not on the card, do the following:

```
}

public Enumeration getClasses(CardType type)
{
     return services.elements();
}
}
```

16.3   The Card Service

Developing the card service is the main topic of this section. The card service has the following purposes:

- *Encapsulation of the card-specific calls from the application.* The application developer does not need to know anything about the CLA and INS bytes and similar things. Therefore the card service should provide an API to the application where the application developer can send and retrieve data in a style well known to programmers. For example, OCF comes with the `FileAccessCardService` card service, which gives an application developer an interface for accessing a file system on the card in a manner similar to reading and writing files on a disk.

- *Interpreting the response to a card command.* Often response data come in a byte stream that is difficult to understand. For example, the response to a GET PROCESSING OPTIONS command cannot easily be processed to determine which data have to be read. Therefore we define helper functions, which return a list of file identifiers and record numbers that can later be read by READ RECORD.

The reader should realize that the card service does not make OCF completely transparent to the application development. Some commands must still be issued by the application (startup and shutdown of the card). These issues will be described in Chapter 17.

Below we describe details of the implementation. Please note that some parts of the card service will be omitted for simplicity.

```
package emvdemo.CardService;


import opencard.core.service.*;
import opencard.core.terminal.*;
import opencard.opt.service.*;
import opencard.opt.util.*;


public class EMVDemoCardService extends CardService
```

This extends the abstract `CardService` class. We must overwrite the `initialize` method. A card channel has to be allocated to provide communication with the card terminal object. Inside `allocateCardChannel` the card channel is obtained from the `CardServiceScheduler` object. In some cases the scheduler can die, especially when the card is removed from the reader. In this case the card channel is set to `null` as well. The next smart card command will therefore return a `CardTerminalException`. Because there is no way that the card service can recover by itself, the application must in this case initiate the creation of a new card service object. In our example, recovering after a card removal is not supported by the application (however, that is an application issue).

```
    protected void initialize( CardServiceScheduler
sched, SmartCard card, boolean block )
    throws InvalidCardChannelException,
CardServiceException
  {
    super.initialize( sched, card, block );

    allocateCardChannel();
  }
```

The following method sends a SELECT command to the card, which selects a card application. The application is identified by its AID, which is explained in Section 16.3. When a response other than 90 00 H is received, a `CardServiceUnexpectedResponse` is thrown, which means that the application is not present in the card. (We assume that we have made no mistakes in building the APDU in a well-tested card service.) Because our card service factory has already tested that the application is present, this is impossible.

16.3.1 SELECT Command

SELECT is a command that switches to our application (in our example, it can be used to select any file):

```
public void Select(byte[] aid,byte length)
    throws
CardTerminalException,CardServiceUnexpectedResponseExce
ption
  {

    CommandAPDU Application=new
CommandAPDU(TransactionConstants.SC_CMD_HDR_LEN+
TransactionConstants.AID_LEN+1);
    Application.append(TransactionConstants.CLA_ISO);
    Application.append(TransactionConstants.INS_SELECT);
    Application.append(TransactionConstants.P1_SELECT);
    Application.append(TransactionConstants.P2_SELECT);
    Application.append(length);
    Application.append(aid);
    Application.append(TransactionConstants.LE_SELECT);
```

Now that we have put together the APDU that will be sent to the card, we can pass it to the card channel, to the card terminal, and further to the reader and the card.

```
    ResponseAPDU
result=getCardChannel().sendCommandAPDU(Application);

    int sw=result.sw();
    if (sw!=TransactionConstants.SW_OK)
    {
      throw new CardServiceUnexpectedResponseException
        ((String)TransactionConstants.SW_SELECT.get(new
Integer(sw)));
    }
  }
```

To make it easier for the application developer, we insert a method that selects our application:

```
public void Select_EMV_Transaction()
    throws
CardServiceUnexpectedResponseException,CardTerminalExce
ption
    {
Select(TransactionConstants.TRANSACTION_AID,Transaction
Constants.AID_LEN);
    }
```

Next, the VERIFY command will be implemented. OCF offers a mechanism to send a "verified APDU," which means one can send an arbitrary command preceded by a VERIFY command. We do not use this, but instead issue our own verification command. Unfortunately this will not work with a reader that has its own PIN pad.

16.3.2 VERIFY Command

The VERIFY command verifies a PIN that has been entered. After verification, the card owner is authenticated to use the card.

```
public int Verify(String PIN)
    throws
CardTerminalException,CardServiceException,CardServiceU
nexpectedResponseException
    {
    byte append_byte;
```

As with every API (or anything that can be accessed by another party), basic consistency checks should occur.[3]

---

3. Many bugs known today, especially security flaws, would not have occurred if consistency checking (e.g., buffer overflows in programs written in C) had been done correctly.

```
if (PIN.length()<4 || PIN.length()>12)
{
throw new CardServiceException("PIN length out of
range (4-12)");
}
CommandAPDU pin=new
CommandAPDU(TransactionConstants.SC_CMD_HDR_LEN+
TransactionConstants.LC_VERIFY);
    pin.append(TransactionConstants.CLA_ISO);
    pin.append(TransactionConstants.INS_VERIFY);
    pin.append(TransactionConstants.P1_VERIFY);
    pin.append(TransactionConstants.P2_VERIFY);
    pin.append(TransactionConstants.LC_VERIFY);
```

Although we have a PIN type other than string, we must convert the PIN. The PIN type is explained in Chapter 15. The conversion process checks whether the PIN is numeric and then packs it into our desired format [2]:

```
pin.append((byte)(TransactionConstants.VERIFY_CONTROL_F
IELD+PIN.length()));
pin.append((byte)0xFF);
    byte[] PIN_Bytes=PIN.getBytes("ASCII");

    append_byte=0;

    for (int i=0;i<12;i++)
    {
      if (i<PIN.length())
      {
        if (PIN_Bytes[i]<'0'|| PIN_Bytes[i]'9')
        {
          throw new CardServiceException("PIN not
          numeric");
        }
        append_byte|=(byte)(PIN_Bytes[i]- 0 );
      }
      else
      {
        append_byte|=(byte)0x0F;
      }
      if (i%2==0)
      {
```

```
    append_byte<<=4;
  }
  else
  {
    pin.append(append_byte);
    append_byte=0;
  }
}
pin.append((byte)0xFF);

ResponseAPDU
result=getCardChannel().sendCommandAPDU(pin);
```

VERIFY is an exception to our usual response treatment. A status word indicating an incorrect PIN does not throw an exception, because EMV requires processing to continue even if the PIN is wrong. For example, the status word 69 83 H indicates a blocked PIN (three or more successive entries of incorrect PINs).

```
    int sw=result.sw();
    if (sw!=TransactionConstants.SW_OK && sw!=0x6983 &&
result.sw1()!=TransactionConstants.WRONG_PIN)
    {
        throw new CardServiceUnexpectedResponseException
        ((String)TransactionConstants.SW_VERIFY.get(new
Integer(sw)));
    }
    return(sw);
}
```

16.3.3 READ RECORD Command

The READ RECORD command reads a specific record out of an AEF. The record is read as a whole. The content of the record is a single TLV object that is, in our application, part of the AFL. The record is identified by the identifier of the AEF (it is an SFI that implicitly selects the AEF) and the record number.[4]

---

4. Note that we had to modify OCF's `opencard.opt.util.TLV` object to allow recursive searching later. The object is derived from `opencard.opt.util.TLV` and is called `ConstructedTLV`.

```
public ConstructedTLV Read_Record(byte SFI, byte rec_nr)
    throws
CardTerminalException,CardServiceException,CardServiceU
nexpectedResponseException
```

As usual, we do some checking to see if the parameter values are reasonable:

```
{
    if (SFI<1 || SFI>31)
    {
        throw new CardServiceException("SFI out of
        range (1-31)");
    }

    CommandAPDU Record=new
    CommandAPDU(TransactionConstants.SC_CMD_HDR_LEN);
    Record.append(TransactionConstants.CLA_ISO);
    Record.append(TransactionConstants.INS_READ_RECORD);
    Record.append(rec_nr);
    Record.append((byte)((SFI<<3)|0x04));
```

As we can see, the SFI fills the five most significant bits of P2. The last 3 bits determine the access mode. The value 4 means that the record is the absolute record specified in P1. Other values (relative to the current selected, first, last, etc.) are not supported.

After this an L_e of 0 is appended (remember—the whole record), the APDU is sent to the card, and the status word is interpreted. The result, a TLV object, is created and returned.

16.3.4   GET PROCESSING OPTIONS Command

This is the first card command not according to ISO 7618. GET PROCESSING OPTIONS returns the location where the application-specific data are stored. The data are returned in the form of a constructed TLV object that contains two TLV objects, one representing the AIP, the other representing the AFL. The AIP is a 2-byte value that indicates which authentication mechanisms the application supports (cardholder verification, issuer authentication, or offline authentication) and whether the terminal risk management task has to be performed. The AFL is a variable-length value that indicates which AEFs and which records store the data needed to perform the transaction. These data will be read by subsequent READ RECORD commands. The format of the AFL cannot easily be interpreted, so we implement some methods that will help interpret it and put it into an easily understandable format.

This command can have more than one result. Several AFLs can be returned, depending on the value of the PDOL, which represents the command data. For our application, we only need one.

Implementation of GET PROCESSING OPTIONS is relatively straightforward in relation to the other commands.

```
public void Get_Processing_Options()
    throws
CardTerminalException,CardServiceUnexpectedResponseExcept
ion
  {
    CommandAPDU proc_opts=new
CommandAPDU(TransactionConstants.SC_CMD_HDR_LEN+
TransactionConstants.LC_GET_PROCESSING_OPTIONS+1);

proc_opts.append(TransactionConstants.CLA_MANUFACTURER);

proc_opts.append(TransactionConstants.INS_GET_PROCESSIN
G_OPTIONS);

proc_opts.append(TransactionConstants.P1_GET_PROCESSING
_OPTIONS);

proc_opts.append(TransactionConstants.P2_GET_PROCESSING
_OPTIONS);

proc_opts.append(TransactionConstants.LC_GET_PROCESSING
_OPTIONS);
    proc_opts.append(TransactionConstants.PDOL_TAG);
    proc_opts.append(TransactionConstants.PDOL_LENGTH);

proc_opts.append(TransactionConstants.LE_GET_PROCESSING
_OPTIONS);

    ResponseAPDU
result=getCardChannel().sendCommandAPDU(proc_opts);

    int sw=result.sw();
    if (sw!=TransactionConstants.SW_OK)
    {
      processing_options=null;
```

```
        throw new CardServiceUnexpectedResponseException
((String)TransactionConstants.SW_GET_PROCESSING_OPTIONS
.get(new Integer(sw)));
        }
    processing_options=new TLV(result.data());
  }
```

As mentioned above, the result is not delivered back, but is instead stored internally. Certain functions are used to retrieve the semantics of the data in a clearer way. First, we use a function that returns the AIP as a number:

```
public int Get_AIP()
   {
      if (processing_options!=null)
      {
          return processing_options.findTag(new Tag
(TransactionConstants.TAG_AIP),null).valueAsNumber();
      }
      else
      {
          return (-1);
      }
   }
```

The next two functions handle the AFL. The AFL is a list of quadruples of bytes, each consisting of:

- A byte containing the SFI where the data are stored. The value is shifted 3 bits to the left. This is not a problem, because an SFI can only take values from 1 to 30. So it could be used as direct input to P2 of the READ RECORD command (except the access mode has to be added). In the interests of clarity for the application developer, we convert it to the normalized value.
- The record number where the data starts.
- The record number where the data ends.
- The number of records, beginning from the start record, that will be included in offline data authentication (static or dynamic).

The first function is used to calculate how many records have to be read. This is useful for dimensioning the array length of the arrays provided as parameters to the second function.

```
public int Get_AFL_Nr_Entries()
  {
    if (processing_options!=null)
    {
        byte[] AFL = processing_options.findTag(new
Tag(TransactionConstants.TAG_AFL),null).valueAsByteArray();
        int len = processing_options.findTag(new
Tag(TransactionConstants.TAG_AFL),null).length();

        int nr=0;
        for(int i=0;i<len i+=4)
        {
            nr+=AFL[i+2]-AFL[i+1]+1;
```

Note that this is a from-to inclusive list, so the formula is last − first + 1.

```
        }
        return(nr);
    }
    else
    {
        return(0);
    }
}
```

The second function returns an array of SFIs and an array of record numbers as specified in the READ RECORD command. The arrays are handed over as parameters to the function.

```
public void Get_AFL(byte[] SFI, byte[] rec_nr)
{
    if (processing_options!=null)
    {
        byte[] AFL = processing_options.findTag(new Tag
(TransactionConstants.TAG_AFL),null).valueAsByteArray();
        int len = processing_options.findTag(new Tag
(TransactionConstants.TAG_AFL),null).length();

        int rec=0;
        for(int i=0;i<len;i+=4)
        {
            for(int j=AFL[i+1];j<=AFL[i+2];j++)
            {
```

```
               SFI[rec]=(byte)(AFL[i]>>3);
               rec_nr[rec]=(byte)j;
               rec++;
           }
        }
     }
  }
```

16.3.5 GENERATE AC Command

Finally, this command will carry out the remaining authentication and
authorization process. After the terminal risk management task has been performed, this command provides all data for the card to perform its card risk
management. Data given are a CDOL, which is a simple concatenation
of data objects with fixed length, because there are no mandatory TLV
structures.

The function has two parameters. One concerns the type of transaction, for example, a transaction to determine whether the terminal is willing
to do a transaction.[5] The second is the transaction data. The function is
called `First_Generate_AC`, because in some cases (ARQC response) a
second GENERATE AC command has to be issued. Because we never
request an ARQC (we never go online), only the first GENERATE AC command is implemented.

Data returned contain a TLV structure that indicates if the card is willing to proceed and a cryptogram that signs the data given to the command
with a key derived from a preshared master key.

```
public TLV First_Generate_AC(int type, ApplicationData
data)
     throws
CardTerminalException,CardServiceException,CardServiceU
nexpectedResponseException

{
     CommandAPDU ac=new
CommandAPDU(TransactionConstants.SC_CMD_HDR_LEN+
TransactionConstants.LC_FIRST_GENERATE_AC+1);
     ac.append(TransactionConstants.CLA_MANUFACTURER);
```

---

5. Even in the event that the terminal is not willing to do a transaction (wrong PIN, card
   expired, etc.), the GENERATE AC command has to be called to notify the card.

```
ac.append(TransactionConstants.INS_GENERATE_AC);
switch (type)
{
    case AAC:
        ac.append(TransactionConstants.
        P1_GENERATE_AC_AAC);
        break;
    case TC:
    ac.append(TransactionConstants.P1_GENERATE_AC_TC);
        break;
default:
        throw new CardServiceException("Illegal
        transaction type");
```

The type of transaction flows in the P1 field of the card command. Values other than AAC and TC throw an exception.

```
}
  ac.append(TransactionConstants.P2_GENERATE_AC);
  ac.append(TransactionConstants.LC_FIRST_GENERATE_AC);

ac.append(data.make_AC1());
```

Object `Application_Data` has a number of bean functions to set and read out the data elements as well as one function that concatenates them to the CDOL-related data.

```
ac.append(TransactionConstants.LE_GENERATE_AC);

    ResponseAPDU
result=getCardChannel().sendCommandAPDU(ac);

    int sw=result.sw();
    if (sw!=TransactionConstants.SW_OK)
    {
        application_cryptogram=null;
        throw new CardServiceUnexpectedResponseException
((String)TransactionConstants.SW_GENERATE_AC.get(new
Integer(sw)));
    }

    application_cryptogram = new TLV(result.data());
```

The result is put into a TLV structure and returned:

```
        return(application_cryptogram);
    }
```

References

[1]   OpenCard Framework 1.2 Programmer's Guide; available at http://www.open-card.org/.

[2]   Europay International S.A., "Off-the-Shelf Card Profile," Oct. 1999.

17

Terminal Application

This chapter is about developing an application on top of the card service. The application will concatenate several calls to card service functions to combine them into a useful transaction. This is the more semantic part of the system and is intended for application programmers who do not need to know exactly how card communication works. For the example we use a part of the EMV specification that will be deployed as a standard for performing all types of debit and credit transactions (Maestro, Cirrus, MasterCard, Visa, or simple ATM[1]) using smart card technology.

17.1 Startup and Shutdown

As with other packages, there are certain procedures for invoking the application and for ending it properly. In the case of the OCF, an overall management class called SmartCard has the member functions start and shutdown (among many others). These functions are static, which means they do not belong to any instance of a SmartCard object. The method start first reads the properties file and retrieves the card terminal factories and the card service factories via the respective registries.

The method shutdown releases all of those factories and closes all card schedulers so that the port(s) to which the reader(s) are allocated are freed for further use. Therefore, shutdown should be called in every possible

---

1. Simple in terms of use, not technical process. ATM transactions are processed via smart cards in many European countries.

case where the application will terminate. Hard termination of the application (like the `kill -9` command under UNIX) can leave the attached reader in an unusable state because there are still locks on the ports. Therefore, the "core" backbone of an OCF application will always look similar to this:

```
try
    {
        SmartCard.start();
        ... // do your card application

    }
    catch (Exception e)
    {
        e.printStackTrace(System.err);
    }
    finally
    {
        try
        {
            SmartCard.shutdown();
        }
        catch (Exception e)
        {
            e.printStackTrace(System.err);
        }
    }
}
```

This does nothing with the card; it merely creates and initializes some objects. To communicate with the card, another static member function of the class SmartCard has to be invoked. This function is called waitForCard. It needs a CardRequest object as a parameter that determines the kind of request. It has to perform three steps:

1. Determine whether to use a card that is already inserted or wait for a new card to be inserted.
2. Define the card terminal object that the resulting object has to support.
3. Define the card service object that the resulting object has to support.

The latter two steps can be NULL to support all terminals/services. The result of waitForCard is a SmartCard object that represents all of the

OCF-related items that are needed to communicate successfully with the card. These include an instance of the card service, the card terminal, and the card scheduler as a connection between the first two. Afterward, the transaction can be performed by obtaining the card service object as the entry point to card communication. Thus the main routine of our application looks like this:

```
public static void main(String[] args)
  {
    IAIK.addAsProvider();
    Locale.setDefault(AUSTRIA);
    TimeZone.setDefault(TimeZone.getTimeZone("CET"));
    try
    {
        SmartCard.start();
        CardRequest c_req=new
CardRequest(CardRequest.ANYCARD,null,null);

        smart_card=SmartCard.waitForCard(c_req);

        trans_svc= (EMVDemoCardService)
smart_card.getCardService(EMVDemoCardService.class,
true);

        Do_Transaction();

    }
    catch (Exception e)
    {
        e.printStackTrace(System.err);
    }
    finally
    {
        try
        {
          SmartCard.shutdown();
        }
        catch (Exception e)
        {
            e.printStackTrace(System.err);
        }
    }
  }
```

Some non-OCF-related initializations occur at the beginning. Because we will later use calendar objects to determine if the card is still valid, we must set the locale and the time zone. The crypto provider must also be initialized. We use the IAIK crypto library (IAIK-JCE v 2.61), which was generously supplied by the Institute for Applied Information Processing and Communications of the University of Technology in Graz, Austria, to develop the application in this book.[2] Of course, other crypto libraries can be used.

17.2   Processing Options and Restrictions

The first part of the application consists of reading data from the card and deciding if the information provided by it is sufficient and valid. Insufficient data lead to an abortion of the transaction. Invalid data, however, lead to a continuation, but with several parameters set such that no successful transaction will be performed; in addition, the various transaction counters will be updated. Here is the code for the first part:

```
trans_svc.Select_EMV_Transaction();

trans_svc.Get_Processing_Options();
```

These function calls do exactly what they were described to do in Chapter 16. They select our application and retrieve the location of the transaction-relevant data:

```
if (trans_svc.Get_AIP()!=AppConstants.AIP)
{
     throw new EMVDemoAppException("Wrong AIP");
}
```

We only support one AIP, so if the AIP retrieved is not ours, abort.

```
int AFL_Nr=trans_svc.Get_AFL_Nr_Entries();
byte[] SFIs=new byte[AFL_Nr];
byte[] recs=new byte[AFL_Nr];

trans_svc.Get_AFL(SFIs,recs);
```

---

2. IAIK offers free licenses for educational purposes; for commercial purposes, license fees must be paid. For further information, see http://jcewww.iaik.tu-graz.ac.at/.

```
    AFL_Data=new ConstructedTLV(new
Tag(0x10,(byte)1,true),trans_svc.Read_Record(SFIs[0],re
cs[0]));
    for(int i=1;i<AFL_Nr;i++)
    {
        AFL_Data.add(trans_svc.Read_Record(SFIs[i],
recs[i]));
    }
```

We initialize our AFL_Data object with the first record retrieved and then successively add all the other records. After that we do the processing restrictions, using this line of code:

```
Processing_Restrictions();
```

The processing restrictions will check the data provided by the card. In our application the following checks will be performed:

- The system checks the AVN.

- The AUC is checked. This decides which kinds of products the card supports (and the terminal is willing to provide the transaction for). There is a distinction between domestic[3] and international transactions, and one in the type of service (EVM defines cash, goods, services, ATM, and cashback).

- If, for example, we only support domestic goods transactions, we need to know where the card was issued, so the country code will be relevant.

- The actual date must not be before the date the card application was activated.[4]

- The card application must not be beyond the expiration date.

- The cardholder verification method must be supported.

The following tests consist of two parts each. First, we check whether a specific tag was delivered by the GET PROCESSING OPTIONS. If not, an exception is thrown. Second, the validity of the value of the tag is tested. If it

3. We use the term *domestic* to mean "in the same country," not inside the United States and Canada.
4. The year in EMV is coded in two digits.

is invalid, a flag is set and terminal risk management will fail. Because the constructed TLV structure is relatively small, there is no performance reason to assign the concrete TLV object to a separate variable, so this will not be done.

```
private static void Processing_Restrictions()
    throws EMVDemoAppException
    {
        if (AFL_Data.findTag(AppConstants.TAG_AVN,
null)==null)
        {
            throw new EMVDemoAppException("No
Application Version Number in AFL");
        }
```

Everything is first checked to see if it exists in the constructed TLV object built by the successive READ RECORD commands.

```
if (AFL_Data.findTag(AppConstants.TAG_AVN,null).
valueAsNumber()!=AppConstants.EMV_AVN)
    {
        VN_wrong=true;
    }
```

Afterward, the version number from the card is compared to the version number of the terminal application. If they are not equal, no successful transaction will occur. This is noted with a flag and will flow into the terminal risk management process.

```
if (AFL_Data.findTag(AppConstants.TAG_AUC,null)==
null)
    {
        throw new EMVDemoAppException("No Application
Usage Control in AFL");
    }

if (AFL_Data.findTag(AppConstants.TAG_COUNTRY_CODE,
null)==null)
    {
        throw new EMVDemoAppException("No Country Code
in AFL");
    }
```

For use of the right application, two parameters need to be evaluated, the application usage control and the country code, because we only allow domestic goods transactions. (This is only for our example; each application must specify what it wants to allow or disallow.)

```
    int auc=AFL_Data.findTag(AppConstants.TAG_AUC,null).
valueAsNumber();
    int country_code=AFL_Data.findTag(AppConstants.TAG_
COUNTRY_CODE,null).valueAsNumber();

    if (((auc&AppConstants.AUC_BIT_DOMESTIC_GOODS)==0)
|| (country_code!=AppConstants.COUNTRY_CODE_AT))
    {
        svc_unavail=true;
    }
```

As mentioned above, in this example we support only domestic goods transactions in Austria. Now validity will be checked. For this we use the GregorianCalendar object from the Java API.[5] There are two time stamps, the application effective date (the date when the application is made valid) and the application expiration date (the date after which the application is no longer valid). Note that the effective date is optional, so if it is omitted, this will not cause an unsuccessful transaction. It simply means that the application has been valid since shortly before the last Ice Age.

```
    byte aefd[];
    GregorianCalendar aefd_cal=new GregorianCalendar();

    boolean no_aefd=false;

    if (AFL_Data.findTag(AppConstants.TAG_AEFD,null)==
null)
    {
        no_aefd=true;
    }
    else
    {

aefd=AFL_Data.findTag(AppConstants.TAG_AEFD,null).value
AsByteArray();
```

---

5. This might seem overengineered, but it is the smartest way to do date comparisons.

```
aefd_cal=new GregorianCalendar(aefd[0]+2000,
                                   aefd[1],
                                   aefd[2]);
}


    if (AFL_Data.findTag(AppConstants.TAG_AEXD,null)
== null)
    {
        throw new EMVDemoAppException("No Application
Expiration Date in AFL");
    }

    byte[]
aexd=AFL_Data.findTag(AppConstants.TAG_AEXD,null).value
AsByteArray();
    GregorianCalendar aexd_cal=new GregorianCalendar
(aexd[0]+2000, aexd[1], aexd[2]);
    GregorianCalendar now=(GregorianCalendar)Calendar.
getInstance();
```

Because we have now instantiated all of our calendar objects, we can do comparisons. Chances are small that we will find applications issued before January 1, 2000, but for the sake of completeness we will define all applications with issuing dates between the years 96 and 99 as twentieth-century issued.[6] We chose 1996 because this was the year when EMV was first proposed.

```
    if (!no_aefd)
    {
        if (aefd_cal.get(Calendar.YEAR)<96&&
now.before(aefd_cal))   // Y2K greets
        {
            app_neff=true;
        }
    }

    if (now.after(aexd_cal))
    {
        app_exp=true;
    }
```

---

6. Of course, 2000 is also part of the twentieth century, but for arithmetic calculations it does not cause problems.

Last but not least, the cardholder verification method is checked. The value of the TLV object contains several fields, two trigger amounts called X and Y, the method, and a value that describes how to use the aforementioned amounts. We use no trigger amounts and only support offline PIN verification (in our example, we wouldn't know where to go online).

```
     if (AFL_Data.findTag(AppConstants.TAG_CVM,null)==
null)
     {
          throw new EMVDemoAppException("No CVM List in
AFL");
     }

     byte[] cvm=AFL_Data.findTag(AppConstants.TAG_CVM,
null).valueAsByteArray();
```

This is the only "true-good" condition, that is, if (and only if) the expression evaluates to true will the transaction be allowed to continue.

```
     for (int i=8;i<AFL_Data.findTag(AppConstants.TAG_
CVM,null).length ();i+=2)
     {
          if (cvm[i]==AppConstants.CVM_OFFLINE_PIN)
          {
               cvm_supp=true;
          }
     }

}
```

After completing the processing restrictions, we proceed to dynamic data authentication, which verifies that the card was issued by the system.

17.3 Dynamic Data Authentication

17.3.1 Certificate Chain

Dynamic data authentication is performed via the card command INTERNAL AUTHENTICATE. This command signs some data with the card private key. The terminal must then verify the signed data with the card public key to prove that the card is not forged. But where does the terminal get this public key? The answer is from certificates. An "agreed-on" certification authority (CA) signs the public key of the issuer and the issuer signs the card key. Thus with the built-in CA public key, the validity of the card public key can be shown. This is called a three-layer public key certification scheme and is illustrated in Figure 17.1.

As we can see, three things must be stored inside the card and one in the terminal application. First the CA public key is stored inside the terminal application and is trusted by it (through some other agreement). This can be used to verify the issuer certificate provided from the card and retrieve the issuer public key. This key can be used to verify the ICC certificate and to retrieve the public key of the ICC. With that the terminal application can verify signatures provided by the ICC and trust the authenticity of this key. Beforehand, during personalization of the ICC, the issuer had to do the following:

- Generate a key pair on the ICC.
- Sign the public key of the ICC with the issuer private key.
- Let the issuer's public key be signed by the trusted CA with its secret key. The CA public key is distributed to the terminals. (Normally a list of trusted CA public keys is provided with the initial setup of the terminal application, as it is in every Web browser.)

A certificate is a public key concatenated in some way with data that uniquely identifies the owner of the public key. It is signed with the private key of the next higher level in the certificate hierarchy. The signature data can be calculated in any of several ways. One is to calculate some hash value and sign the hash value. The hash value is usually relatively short in relation to the key length (160 to 1,024 bits or higher). Therefore padding bits must

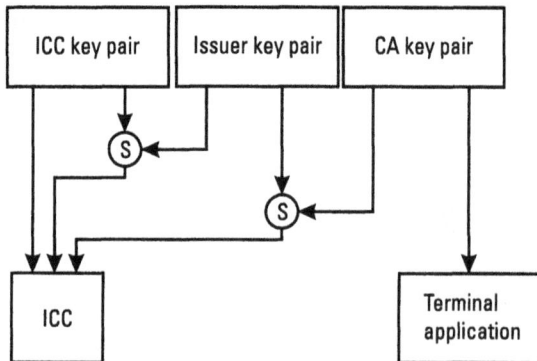

Figure 17.1 Three-layer public key certification scheme.

be used to get the cryptogram length equal to the key length. The padding has to have the most significant bit set to 0 and one of the next bits should be set to 1 (not necessarily the second bit but commonly so). Even a defined trailer byte is a good idea to make it easy to identify corrupted signatures. The rest can be filled with zeroes or random data (as in public key cryptography standard PKCS #1 v1.5 [1]).

In this case the public key and the certificate data have to be stored separately on the card. Because space is always in short supply in smart card applications, this might be waste of storage. So the certificate data and (part of) the key are used as padding data. We say "part of" because the length of the public key may not be much smaller than the key of the signer in which it can fit. One possibility then is only to put in as much data as fits and leave the rest in a separate, nonsigned field. This is how it is done in EMV. For example, let us show the encoding of the issuer certificate (the encoding of the ICC certificate is very similar; for a full explanation, see [2], Section 2.2):

- Header byte (6A H);
- Certificate format (02 H);
- Issuer identification number, which is part of the PAN;
- Certificate expiration date (only month and year);
- Certificate serial number;
- Hash algorithm indicator;
- Issuer public key algorithm indicator;
- Issuer public key length;
- Issuer public key exponent length;
- Issuer public key (or the leftmost 36 bytes of it);
- Hash result;
- Recovered trailer (BC H).

Input to the hash function is the above data (except header and trailer) and, of course, the remainder of the public key (the modulus), if necessary, and the public exponent. The public exponent in EMV is always 3 or F4, where F4 is Fermat's fourth number,[7] which is 65,537. These values are very popular for public key exponents (few 1s in their binary representation).

---

7. Fermat's nth number is $2^{2^n} + 1$.

17.3.2 Signing Authentication Data

Now that we have the card public key and know it belongs to the card, we can do the internal authentication. The terminal generates a random number (4 bytes or more). Together with its ID, country code, and the current date it forms the dynamic data object list (DDOL). The DDOL is sent to the card and the card takes it as input to its signature. In addition to the DDOL-related data, some dynamic data from the ICC flow into signature generation (8 random bytes). The signature and the DDOL are then sent back to the terminal in response to the INTERNAL AUTHENTICATE command. The DDOL is returned to ensure that no transmission errors occurred.

17.4   Terminal Risk Management

The terminal risk management task is the part of the process where the terminal decides if it is willing to process the transaction. Two parts flow into this process. First, the data evaluated in the processing restrictions must be valid, and the data entered by the user must be valid (the PIN entered correctly). The PIN is entered through a standard dialog pop-up window that comes with OCF:

```
DefaultCHVDialog Diag=new DefaultCHVDialog();
    String PIN=Diag.getCHV(0);

    int Verify_sw=0;

    if (PIN.length()>0)
    {
        Verify_sw=trans_svc.Verify(PIN);
    }
    else
    {
        empty_pin=true;
    }
```

Next, a data structure representing the CDOL-related data will be built. Most of the data elements are transaction parameters, which are constant for the sample application. *Terminal verification results* that have to be constructed, however, are one exception. They are assembled from 40 bits whose meaning is defined in [3, pp. 50ff]. The CDOL-related data consists of this information:

- The transaction amount;

- The amount for cashback transactions. If this type of transaction is allowed at all, the amount is likely to be far less than the maximum amount of other transactions;

- The country code of the country issuing the terminal (the value for Austria is 40);

- The terminal verification results;

- The currency code of the currency accepted by the terminal (the Euro has code 978);

- The current date;

- The transaction type. This is actually the first byte of the AUC that determines which transactions are allowed (goods, services, ATM, cash, domestic as well as international; cashback, however, is not handled here);

- A 32-bit unpredictable number;

- The terminal type, distinguished between offline/online, attended/ unattended, and the type of client (cardholder, merchant, financial institution). We use cardholder/unattended/offline, which results in a value of 36 according to [2];

- The data authentication code, a preshared value deployed by the issuers of both card and terminal.

Now the terminal action analysis is done, which means the terminal has decided whether to accept or decline the transaction. This is how the code looks:

```
app_dat=new ApplicationData();
app_dat.setCurrencyMultiplicator(AppConstants.CURR_MULT
IPLICATOR);
    app_dat.setAuthAmount(AppConstants.AMOUNT_AUTH);
    app_dat.setOtherAmount(AppConstants.AMOUNT_OTHER);
```

Amounts are given in minor units of the currency (= cents) so the amounts must be multiplied with the currency multiplier, which is 100 in most, if not all, countries.

```
app_dat.setTerminalCountry(AppConstants.COUNTRY_CODE_AT
);

    byte[] TVR=new byte [5];
    Arrays.fill(TVR,(byte)0);

    if (!dda_auth)
    {
        TVR[0]=AppConstants.BIT_OFF_LINE_DDA_FAILED;
        type=EMVDemoCardService.AAC;
    }
```

The transaction fails if the dynamic data authentication process from Section 17.3 fails. In this (and every other case when something goes wrong), the terminal sets the reference control parameter to AAC, which means that the transaction is declined, but nevertheless the GENERATE AC command is processed. The results from the processing restrictions flow in:

```
    if (VN_wrong)
    {
        TVR[1]|=AppConstants.BIT_VERSION_NUMBER_WRONG;
        type=EMVDemoCardService.AAC;
    }

    if (app_neff)
    {
        TVR[1]|=AppConstants.BIT_APP_NOT_EFFECTIVE;
        type=EMVDemoCardService.AAC;
    }

    if (app_exp)
    {
        TVR[1]|=AppConstants.BIT_APP_EXPIRED;
        type=EMVDemoCardService.AAC;
    }

    if (svc_unavail)
    {
        TVR[1]|=AppConstants.BIT_REQUESTED_SERVICE_NA;
        type=EMVDemoCardService.AAC;
    }
```

Now the PIN parameters flow in. These include the supplied card verification method as well as the VERIFY result. Because the card knows if it has verified its PIN successfully before (it has a built-in state machine), this might seem obsolete, but it is needed and moreover useful for dumping card communications to, for instance, a journal file.

```
if (Verify_sw>>8==TransactionConstants.WRONG_PIN)
{
    TVR[2]|=AppConstants.BIT_CHV_FAILED;
    type=EMVDemoCardService.AAC;
}

if (!cvm_supp)
{
    TVR[2]|=AppConstants.BIT_CVM_UNAVAIL;
    type=EMVDemoCardService.AAC;
}

if (Verify_sw==0x6983)
{
    TVR[2]|=AppConstants.BIT_PIN_COUNTER_BLOCKED;
    type=EMVDemoCardService.AAC;
}

if (empty_pin)
{
    TVR[2]|=AppConstants.BIT_PIN_EMPTY;
    type=EMVDemoCardService.AAC;
}

TVR[3]=AppConstants.BIT_DEF_TDOL;

app_dat.setTerminalVerificationResults(TVR);
```

All terminal verification result bits that our application is able to set are now set. We therefore fill out the remaining CDOL-related fields:

```
app_dat.setTransactionCurrency(AppConstants.CURR_CODE_E
URO);
    app_dat.setTransactionDate(Calendar.getInstance());
    app_dat.setTransactionType(AppConstants.TT_GOODS);
```

```
    byte [] rnd=new byte[]{ (byte)0x01, (byte)0x33,
(byte)0x02, (byte)0x11} ;
    byte [] rnd=new byte[ApplicationData.AC_RANDOM/2];
    rand.nextBytes(rnd);
    app_dat.setUnpredictableNumber(rnd);
    app_dat.setTerminalType(AppConstants.OFFLINE_TERMINAL);
    app_dat.setDAC(AppConstants.DAC);
```

Finally, the GENERATE AC command can be issued. As mentioned in Chapter 16, there can be two GENERATE AC Commands, but because we never go online, the second is never issued.

```
    TLV AC_1=trans_svc.First_Generate_AC(type,app_dat);
    if (Verify_Response(AC_1)==AppConstants.TC)
    {
        System.out.println("SUCCESS");
    }
    else
    {
        System.out.println("FAILURE");
    }
```

After the GENERATE AC command is issued, additional script processing can be performed. For simplicity's sake, the verification, which is normally performed by the issuer, is done by our application.

17.5   Verification of the Result

17.5.1 General Verification

This part covers general verification issues such as whether the card is willing to do the transaction, whether we have the same DACs on both sides, and even whether the ACs are equal. AC calculation itself is explained in Section 17.5.2.

First we must see if the data returned from GENERATE AC contain all TLV objects (the normal case):

```
if (AC.findTag(AppConstants.TAG_CID,null)==null)
    {
        throw new EMVDemoAppException("No CID in AC
Response");
    }
```

```
if (AC.findTag(AppConstants.TAG_ATC,null)==null)
{
      throw new EMVDemoAppException("No Application
Transaction Counter in AC Response");
}
if (AC.findTag(AppConstants.TAG_IAD,null)==null)
{
      throw new EMVDemoAppException("No Issuer
Application Data in AC Response");
}
if (AC.findTag(AppConstants.TAG_ARQC_VAL,null)==
null)
{
      throw new EMVDemoAppException("No ARQC Val in
AC Response");
}
```

Next, we retrieve the cryptogram information data (CID) from the response. This can be one of three possibilities: TC (accept), AAC (decline), or ARQC (go online). We do not go online at all in our application, so ARQC is treated as AAC.

```
int CID=AC.findTag(AppConstants.TAG_CID,null).value
AsNumber();

if (CID==AppConstants.ARQC)
{
      CID=AppConstants.AAC;
}
```

We construct the input array for AC calculation. The constants from object `ApplicationData` are halved because this helper object administrates the field lengths in nibbles. The input array contains all but the last two fields (terminal type and DAC) provided as data to the GENERATE AC command plus the AIP plus the ATC plus the CVR, which is part of the issuer application data (IAD) (bytes 3 to 6).

```
System.arraycopy(app_dat.make_AC1(),0,ARQC_val,0,app_dat.
AC1_SUM/2);
      ARQC_len=ApplicationData.AC1_SUM/2-ApplicationData.
AC_TYPE/2-ApplicationData.AC_DAC/2;
```

```
        ARQC_val[ARQC_len]=(byte)(AppConstants.AIP>>8);
        ARQC_val[ARQC_len+1]=(byte)(AppConstants.AIP & 0xff);
        ARQC_len+=AppConstants.AIP_LEN;

System.arraycopy(AC.findTag(AppConstants.TAG_ATC,null).
valueAsByteArray(),0,ARQC_val,ARQC_len,AC.findTag(App-
Constants.TAG_ATC,null).length());

ARQC_len+=AC.findTag(AppConstants.TAG_ATC,null).length();

System.arraycopy(AC.findTag(AppConstants.TAG_IAD,null).
valueAsByteArray(),2,ARQC_val,ARQC_len,AC.findTag(App-
Constants.TAG_IAD,null).length()-4);
ARQC_len+=AC.findTag(AppConstants.TAG_IAD,null).len-
gth()-4;
```

Then the AC is calculated. The result is compared to the value the card has calculated, and if they are not the same, the transaction fails:

```
        byte[]
ARQC_AC=Calc_AC(ARQC_val,ARQC_len,AC.findTag(AppConstan
ts.TAG_ATC,null).valueAsByteArray(),app_dat.getUnpredic
tableNumber());
        if (!Arrays.equals(ARQC_AC,AC.findTag(AppConstants.
TAG_ARQC_VAL ,null).valueAsByteArray()))
        {
                CID=AppConstants.AAC;
        }
```

As the last step, the DACs are compared. As mentioned before, the DAC is a cryptogram precalculated by the card/terminal issuer and entered into the devices during personalization. For our example we can simply assume this is an agreed-on constant. It is retrieved from the last two bytes of the IAD:

```
        byte[] DAC=new byte[2];

System.arraycopy(AC.findTag(AppConstants.TAG_IAD,null).
valueAsByteArray(),AC.findTag(AppConstants.TAG_IAD,null
).length()-2,DAC,0,2);

        if (!Arrays.equals(DAC,app_dat.getDAC()))
        {
```

```
        CID=AppConstants.AAC;
    }

    return CID;
```

To summarize, if everything goes well, the TC value survives all tests and is returned. If something goes wrong, AAC is returned.

17.5.2 AC Calculation

AC calculation is done using two DES keys. The AC is calculated using some DES-MAC algorithm. All but the last block of the input to the calculation is encrypted in CBC mode and the result of the exclusive-or operation of the last cleartext block and the last CBC encrypted bock is encrypted in Triple DES. The calculation is not pure Triple DES-EDE but it is secure like Triple DES-EDE. The key used is a derived key from a preshared master key. Key derivation is explained and illustrated in Section 14.2.2.

We do our cryptographic calculation with the help of IAIK's crypto library as mentioned earlier.

```
Cipher derivation_cipher =
Cipher.getInstance("3DES/ECB/NoPadding", "IAIK");

    byte[] key_value=new byte[8];

    System.arraycopy(ATC,0,key_value,0,ATC.length);
    System.arraycopy(UN,0,key_value,4,UN.length);
```

We construct the base of the key derivation from the ATC and the random number submitted by the GENERATE AC command:

```
    Key key = new iaik.security.cipher.SecretKey(App
Constants.Master_Key, "RAW");
    derivation_cipher.init(Cipher.ENCRYPT_MODE,key);

    key_value[2] = (byte)0xF0;
    byte[] Secret_Key_l =
derivation_cipher.doFinal(key_value);
    key_value[2] = (byte)0x0F;
    byte[] Secret_Key_r =
derivation_cipher.doFinal(key_value);
```

We have derived the correct session keys. Thus we have everything that is needed to calculate the AC. First the data has to be padded according to [4]:

```
Cipher cipher =
Cipher.getInstance("DES/CBC/NoPadding", "IAIK");
    byte[] padded=new byte[(length/8+1)*8];

    Arrays.fill(padded,(byte)0);
    System.arraycopy(plaintext,0,padded,0,length);
    padded[length]=(byte)0x80;
```

Then, with the derived keys, the AC is calculated. Here is all but the last block:

```
    Key key_l = new iaik.security.cipher.SecretKey
(Secret_Key_1, "RAW");
    Key key_r = new iaik.security.cipher.SecretKey
(Secret_Key_r, "RAW");

    IvParameterSpec iv = new IvParameterSpec(App
Constants.IV);

    cipher.init(Cipher.ENCRYPT_MODE, key_l, iv);

    byte[] intermediate_1 = cipher.doFinal(padded);
```

For the last block, another cipher has to be created. Though the last block is actually Triple DES-EDE, it is calculated as if it were a concatenation of DES crypto operations. Let's look quickly at how Triple DES-EDE works: First there is an encrypt operation with the first (leftmost) key. Then (for historical reasons) there is a decrypt[7] operation with the other keys. Then the first encryption procedure is repeated with the first key.

```
    cipher = Cipher.getInstance("DES/ECB/NoPadding",
"IAIK");
    byte[] intermediate_1_last_part=new byte[8];
    System.arraycopy(intermediate_1,intermediate_1.len-
gth-8,intermediate_1_last_part,0,8);
    cipher.init(Cipher.DECRYPT_MODE, key_r);
```

---

7. This was because it was believed that a DES operation had the mathematical attributes of a group. In fact it does not, which was proved by [5].

```
    byte[] intermediate_2 =
cipher.doFinal(intermediate_1_last_part);
    cipher.init(Cipher.ENCRYPT_MODE, key_1);
    byte[] ciphertext=cipher.doFinal(intermediate_2);

    return ciphertext;
```

Whatever this calculation returns it is passed back as an input to result in verification.

References

[1]   RSA Laboratories "PKCS #1 RSA Cryptographic Standard"; available at http://www.rsasecurity.com/rsalabs/pkcs/pkcs-1/.

[2]   Europay International S. A., "Integrated Circuit Card Application Specification for Pay Now (Debit) and Pay Later (Credit) Cards," Oct. 1999.

[3]   Europay International S. A., "Off-the-Shelf Card Profile," Oct. 1999.

[4]   ISO/IEC 9797, "Information Technology—Security Techniques—Data Integrity Mechanism Using a Cryptographic Check Function Employing a Block Cipher Algorithm," 1994 (available in English only).

[5]   Campbell, K. W., and M. J. Wiener. "DES Is Not a Group," *Advances in Cryptology (Crypto '92)*, Berlin: Springer-Verlag, 1993, pp. 512–520.

18

Conclusion

This book presented and discussed a variety of modern technologies related to the e-payment applications of today and tomorrow. We started with smart card technology and explained its essential principles, security characteristics, and application domains. We paid special attention to Java Card technology, a relatively new branch in smart card technology. Any smart card system includes not only a smart card itself but also a card terminal—a computer device capable of communicating with the card and performing certain application-specific tasks. Platform independence and interoperability play an important role in terminal applications. Therefore, we also took a look at the OCF, a Java-based framework that allows the development of platform-independent smart card terminal applications.

As of 2001, far more Java Cards have been issued to end customers than MULTOS cards or Windows for Smart Cards cards, but Java Cards have still not reached their full market potential. Three reasons can be identified. First, the worldwide smart card market has an annual growth rate of about 15%. Second, smart card ICs are becoming more powerful every year, which implies that ICs capable of running applications on top of a Java virtual machine are becoming cheaper. Java cards are presently used only in those fields of application that allow higher card prices, but other fields will produce additional market penetration. One of these new fields of Java Card application will be the banking cards sector. (The debit and credit card example we chose in this book is not a typical application today, because most banking cards only require ICs to have a medium range of performance.) Third, there will be a general demand for shorter development cycles in the smart card world. Java Card helps to reduce development time significantly because there is no need to manufacture customized ICs for every new application code, which can take several months. ICs with standardized Java Card APIs can be taken from stock, loaded with the new application, and used instantly. Higher volumes will further reduce prices and thus speed up the whole process.

Some new technical developments are expected in the early 2000s. Semiconductor manufacturers have already presented ICs that can execute Java byte code directly at the hardware level. Today this merely assists the card operating systems in executing the most frequent commands more efficiently. However, manufacturers will soon be able to deliver a fully Java Card–compliant card OS with the hardware.

Another expected development concerns interoperability. Some of the different implementations of Java Card available in 2001 differ in some details, for example, the cryptographic packages, the application loading mechanism, and other initialization and personalization functions. Additional specifications to Java Card that ensure true interoperability have to be defined in order to achieve the goal that was set at the beginning: that the same application could be loaded from the same development kit to different Java Card types. One can see that this process has already begun in specifications like Visa Open Platform or GSM Java Card class packages (see Chapter 11).

Security evaluations are also becoming more and more important in the smart card world. Because Java Card is a multiapplication platform, there will be a strong requirement from the market for security evaluations of the operating system, the JCVM, and the framework class packages. It is to be hoped that one day we can load our applications onto interchangeable Java Card ICs from different vendors, all with proven security and differing only in such matters as memory size, performance, and commercial conditions.

A set of EMV specifications is the result of joint efforts by Europay, MasterCard, and Visa to make possible cross-banking credit/debit smartcard-based payment applications that can be used worldwide. We chose EMV specifications as a basis for our sample payment application and explained the EMV transaction flow in detail.

The best way to understand a technology is to try it. That is exactly the principle we followed in explaining how to develop credit/debit applications using Java Card and OCF. In the book, we not only gave an abstract description of smart cards and Java Card, OCF, and EMV but also presented a real and running sample application developed for Java Card in accordance with the EMV specifications.

Our goal was to demonstrate the principles of EMV credit/debit applications. Therefore, our sample application does not implement all functions defined in the EMV specifications and is not a ready-to-use real-life EMV application. Our focus was on the main steps of the EMV transaction flow and transaction-related parameters, security-related functions, and decision-making routines.

For simplicity's sake, we omitted card terminal authentication as well as online transaction processing. Although those two functions are indeed an integral part of any real-life EMV application, their absence in our sample application is not detrimental to understanding the principles of EMV application implementation on the Java Card and OCF platforms and may even make the sample easier to follow.

We would like to draw attention to one more aspect of our sample application—separation of functionality. In both parts of our sample application, that is, in the terminal application and in the card application, smart card communication and EMV payment functions are clearly separated. In the card application, APDU handling is performed by the card applet class and EMV transaction processing is performed by another class containing all EMV-related functionality. In the terminal application, smart card communication routines are covered by the OCF card service, and the EMV terminal functionality is inherited in a separate class layered on top of the card service.

This approach not only increases the modularity and extendibility of the applications, but also allows an even higher degree of technology independence. In other words, the application could easily run (with minor modifications in the worst case) on further versions of both Java Card and OCF that might introduce certain modifications in API and core classes. Technology development is rapid and, no doubt, Java Card and OCF will be also evolving during the next few decades. Real-life applications must take this aspect into account and be designed in such a way that future modifications of underlying technologies will not make significant modifications of the applications necessary.

While developing the sample application, we always kept interoperability issues in mind. These mainly concern the card application. Unfortunately, the interoperability characteristics of Java Card implementations existing today leave much to be desired—smart card manufacturers frequently introduce their own, proprietary, class packages, which significantly decreases the interoperability of applications developed for one particular type of Java Card.

In the interests of interoperability, we avoided using proprietary classes supplied by our Java Card manufacturer, Giesecke & Devrient (they are, in fact, just some utility classes not influencing the core Java Card API). Therefore, our card application should be able to run on any kind of Java Card 2.1 smart card. The only problem that could arise is related to cryptographic classes that are not supported on some Java Card implementations.

Also the terminal application has an abstraction between the classes provided by Giesecke & Devrient (for accessing their readers) and the classes written for the application; this allows us to grant interoperability for the terminal application with the card readers of other vendors.

Appendix A:
Card Applet Source Code

This appendix contains the complete source code of the sample card application presented in the book

```
/*- - - - - - - - - - - - - - - - - - - - - - - - - - */
/* File EMV.java                                       */
/* Contains definition of application constants        */
/*                                                     */
/*- - - - - - - - - - - - - - - - - - - - - - - - - - */

package emvdemo;

/* Public interface EMV                                */
/* Defines EMV specific constants                      */

/* An abstract interface comprising the EMV specific   */
/* constants                                           */
public abstract interface EMV
{

    // Object tags
    public final static byte TAG_RECORD = (byte) 0x70;
    public final static byte TAG_PROC_OPT = (byte) 0x77;
    public final static byte TAG_RESP = (byte) 0x77;
    public final static byte TAG_AIP = (byte) 0x82;
    public final static byte TAG_AFL = (byte) 0x94;
    public final static byte  TAG_CVM = (byte)  0x8E;
    public final static short  TAG_CHN = (short) 0x5F20;
```

```
public final static short  TAG_ACD = (short) 0x9F42;
public final static short  TAG_ICD = (short) 0x5F28;
public final static short  TAG_AXD = (short) 0x5F24;
public final static short  TAG_AED = (short) 0x5F25;
public final static short  TAG_AUC = (short) 0x9F07;
public final static short  TAG_AVN = (short) 0x9F08;
// Cryptogram information data
public final static short  TAG_CID = (short) 0x9F27;


public final static short  TAG_ATC = (short) 0x9F36;
public final static short  TAG_AC  = (short) 0x9F26;
public final static short  TAG_IAD = (short) 0x9F10;
    //Issue application data


// Length of fixed-length objects
public final static byte LEN_CID = (byte) 0x01;
public final static byte LEN_ATC = (byte) 0x02;
public final static byte LEN_AC  = (byte) 0x08;
public final static byte LEN_IAD = (byte) 0x08;
public final static byte LEN_GAC1 =(byte) 0x1F;


// Offsets in complex TLV objects
// Response to 1 Generate AC
public final static byte OFFSET_CID = (byte) 2;
public final static byte OFFSET_ATC = (byte) 6;
public final static byte OFFSET_AC  = (byte) 11;
public final static byte OFFSET_IAD = (byte) 22;


public final static byte OFFSET_TUN = (byte)25;
// Offset of the Terminal UN on CDOL1


// Coding of P1 in Generate AC
public final static byte  CODE_TC = (byte) 0x40; // TC
public final static byte  CODE_AAC = (byte) 0x00; // AAC


// Flags
// Constants defining the flags location in the flag object SequenceFlag
public final static short SEQF_LENGTH = 7; // Number of flags
public final static byte  SEQF_APP_SELECTED = 0; // Application selected
public final static byte  SEQF_APP_INVALIDATED = 1;
    // Application invalidated
// GET PROCESSING OPTIONS was performed
public final static byte  SEQF_GETPROC_PERFORMED = 2;
// A first GENERATE AC was executed with
// an ARQC response
public final static byte  SEQF_ARQC_GENERATED = 3;
// A first GENERATE AC was executed with
// an AAC response because the application is invalidated
public final static byte  SEQF_AAC_GENERATED = 4;
// The VERIFY command was performed
```

```
public final static byte  SEQF_PIN_PERFORMED = 5;
// The VERIFY command was successfully executed
public final static byte  SEQF_PIN_VERIFIED = 6;

// Class codes
public final static byte CLA_ISO = (byte)0x00;
public final static byte CLA_MANUFACTURER = (byte)0x80;

// Instruction codes
public final static byte INS_SELECT = (byte)0xA4;
public final static byte INS_READ_RECORD = (byte)0xB2;
public final static byte INS_GET_PROCESSING_OPTIONS = (byte)0xA8;
public final static byte INS_VERIFY = (byte)0x20;
public final static byte INS_GENERATE_AC = (byte)0xAE;

// P1 codes
public final static byte P1_SELECT = (byte)0x04;
public final static byte P1_GET_PROC_OPT = (byte)0x00;
    public final static byte P1_VERIFY = (byte)0x00;

// P2 codes
public final static byte P2_SELECT = (byte)0x00;
public final static byte P2_GET_PROC_OPT = (byte)0x00;
public final static byte P2_VERIFY = (byte)0x80;
public final static byte P2_GENERATE_AC = (byte)0x00;

// Lc
public final static byte LC_GET_PROC_OPT = (byte)0x02;
public final static byte LC_VERIFY = (byte)0x08;
public final static byte LC_GAC = (byte)0x20;

//Error status bytes
final static short SW_FILE_INVALID = (short) 0x6283;
    // Selected file invalidated
// Verification failed, 0 Retries
final static short SW_VER_FAILED_0 = (short) 0x63C0;
// Verification failed, 1 retries
final static byte  SW1_VER_FAILED_1 = (short) 0x63C1;
// Verification failed, 2 retries
final static short SW_VER_FAILED_2 = (short) 0x63C2;
final static byte  SW1_VER_FAILED = (byte)  0x63;
// State of non-volatile memory unchanged
final static short SW_MEM_UNCH = (short) 0x6400;
final static short SW_MEM_FAILURE = (short) 0x6581; // Memory failure
final static short SW_WRONG_LEN = (short) 0x6700; // Wrong length
// Conditions of use not satisfied
final static short SW_COND_NOTSAT = (short) 0x6985;
// Command incompatible with file organization
final static short SW_COMM_INCOMP = (short) 0x6981;
// Security status not satisfied
```

```java
final static short SW_SEC_MOTSAT = (short) 0x6982;
// Authentication Method blocked
final static short SW_AUTHM_BLK = (short) 0x6983;
// Reference Data Invalidated
final static short SW_REFD_INVALID = (short) 0x6984;
final static short SW_FUNC_NSUPP = (short) 0x6A81; // Function not supported
final static short SW_FILE_NFOUND = (short) 0x6A82; // File not found
final static short SW_REC_NFOUND = (short) 0x6A83; // Record not found
final static short SW_WRONG_P1P2 = (short) 0x6A86; // Incorrect P1 P2


// Other constants
final static byte  AID_MAX_LEN = 16;
final static byte  AID_MIN_LEN = 5;
final static short ATC_MAX_VALUE = (short) 0xFFFF;
final static byte  PIN_TRY_LIMIT = 3;
final static byte  PIN_MAX_LEN = 8;
final static short AIP_LEN = (short) 2;
final static short AFL_LEN = (short) 8;
final static byte  GAC1_RESP_LEN = (byte) 0x21;
                              // Length of the value field
                              // of the response to the 1


// Generate AC
final static byte MESSAGE_LEN = (byte) 40;
    // Length of the message - input to AC
final static byte AC_LEN = (byte) 8; // Cryptogram length
final static byte SKEY_LEN = (byte) 8; // session key length
final static byte MKEY_LEN = (byte) 16; // master key length


final static byte  MAX_FILE_SIZE = 30; // maximum possible file size
final static byte NUMBER_LEN = 6; // length of the long number in bytes
```

```
/*- - - - - - - - - - - - - - - - - - - - - - - - - - */
/* File EMVdemo.java                                  */
/* Contains implementation of the                    */
/* EMVdemo applet class                               */
/*                                                    */
/*- - - - - - - - - - - - - - - - - - - - - - - - - - */

package emvdemo;

import javacard.framework.*;

public class EMVdemo extends Applet

{

    // EMV applet objects
    private EMVPurse emvPurse; // EMV purse object

        //
        public final static byte TRUE  = 1;
        public final static byte FALSE = 0;

        /* Install method is invoked by the card JCRE on the*/
        /* end of the applet installation procedure         */
        /* In the method, we create the applet instance     */
        public static void install (byte[] buffer, short offset, byte length)
        {
            new EMVdemo();
        }


        /* Private constructor of the Applet                */
        /* We register the applet within JCRE               */
        private EMVdemo()
        {
            emvPurse = new EMVPurse();
            register();
        }

        public boolean select (APDU apdu)
        {
            return true;
        }

        /* Method process processes the revoked by JCRE to process */
        /* the incoming APDU                                       */
        public void process(APDU apdu) throws ISOException
        {
            byte[] buffer = apdu.getBuffer();
            byte[] response;
```

```
byte sfi, rn, rl;
short len;
EMVFileRecord record;

if (selectingApplet())
{
    // Modifying the flags
    emvPurse.sequenceFlag[EMV.SEQF_APP_SELECTED] = TRUE;
    emvPurse.sequenceFlag[EMV.SEQF_GETPROC_PERFORMED] = FALSE;
    emvPurse.sequenceFlag[EMV.SEQF_ARQC_GENERATED] = FALSE;
    emvPurse.sequenceFlag[EMV.SEQF_AAC_GENERATED] = FALSE;
    emvPurse.sequenceFlag[EMV.SEQF_PIN_PERFORMED] = FALSE;
    emvPurse.sequenceFlag[EMV.SEQF_PIN_VERIFIED] = FALSE;

    return;
}
// Let us process the buffer containing the incoming APDU
switch (buffer[ISO7816.OFFSET_CLA])
{
case ISO7816.CLA_ISO:
    switch ( buffer[ISO7816.OFFSET_INS])
    {

    case EMV.INS_SELECT:

    if ( (buffer[ISO7816.OFFSET_LC] EMV.AID_MIN_LEN) ||
        (buffer[ISO7816.OFFSET_LC]  EMV.AID_MAX_LEN) )
        ISOException.throwIt(ISO7816.SW_WRONG_LENGTH);

    if ( (buffer[ISO7816.OFFSET_P1] != EMV.P1_SELECT) ||
        (buffer[ISO7816.OFFSET_P2] != EMV.P2_SELECT) )
        ISOException.throwIt(EMV.SW_WRONG_P1P2);

    break;

    case EMV.INS_READ_RECORD:

    if ( buffer[ISO7816.OFFSET_P1] == 0 )
        ISOException.throwIt(EMV.SW_FUNC_NSUPP);

    // last three bits must be 100 B
    if ( (buffer[ISO7816.OFFSET_P2] & 7) != 4)
        ISOException.throwIt(EMV.SW_FUNC_NSUPP);

    sfi = (byte) (buffer[ISO7816.OFFSET_P2]>>>3);
    rn  = buffer[ISO7816.OFFSET_P1];
```

```
            record = emvPurse.readRecord(sfi,rn);
            rl = record.getActualLen();


            Util.arrayCopy(record.getData(), (short)0,
                       buffer, (short)0, (short) rl);


            apdu.setOutgoingAndSend((short)0, (short)rl);


            break;


            case EMV.INS_VERIFY:
            if ( (buffer[ISO7816.OFFSET_P1] != EMV.P1_VERIFY) ||
                (buffer[ISO7816.OFFSET_P2] != EMV.P2_VERIFY) )
                ISOException.throwIt(EMV.SW_WRONG_P1P2);


            if ( buffer[ISO7816.OFFSET_LC] != EMV.LC_VERIFY )
                ISOException.throwIt(ISO7816.SW_WRONG_LENGTH);


            emvPurse.sequenceFlag[EMV.SEQF_PIN_PERFORMED] = TRUE;
            emvPurse.sequenceFlag[EMV.SEQF_PIN_VERIFIED]  = FALSE;


            if (emvPurse.getPINTriesRemaining() == 0)
                ISOException.throwIt(EMV.SW_AUTHM_BLK);


            apdu.setIncomingAndReceive();


            if ( emvPurse.checkPIN(buffer,(short)ISO7816.OFFSET_CDATA,
                   EMV.PIN_MAX_LEN) )
            {
            emvPurse.sequenceFlag[EMV.SEQF_PIN_VERIFIED]  = TRUE;
            return;
            }
            else
                ISOException.throwIt(
                   Util.makeShort(EMV.SW1_VER_FAILED,
                       emvPurse.getPINTriesRemaining()));
            break;
        default:
            ISOException.throwIt(ISO7816.SW_INS_NOT_SUPPORTED);
        }


    break;


case EMV.CLA_MANUFACTURER:
    switch (buffer[ISO7816.OFFSET_INS])
    {
    case EMV.INS_GET_PROCESSING_OPTIONS:
```

```
        if ( (buffer[ISO7816.OFFSET_P1] != EMV.P1_GET_PROC_OPT) ||
        (buffer[ISO7816.OFFSET_P2] != EMV.P2_GET_PROC_OPT) )
            ISOException.throwIt(EMV.SW_WRONG_P1P2);

        if ( buffer[ISO7816.OFFSET_LC] != EMV.LC_GET_PROC_OPT )
            ISOException.throwIt(ISO7816.SW_WRONG_LENGTH);

if ( emvPurse.sequenceFlag[EMV.SEQF_APP_SELECTED] != TRUE ||
    emvPurse.sequenceFlag[EMV.SEQF_GETPROC_PERFORMED] == TRUE )
        ISOException.throwIt(EMV.SW_COND_NOTSAT);

        if ( emvPurse.getATC() == EMV.ATC_MAX_VALUE )
            ISOException.throwIt(EMV.SW_REFD_INVALID);

        // Incrementing the transaction counter
        emvPurse.incrementATC();

        // Setting the sequence flags
        emvPurse.sequenceFlag[EMV.SEQF_GETPROC_PERFORMED]  = TRUE;
        emvPurse.sequenceFlag[EMV.SEQF_ARQC_GENERATED]     = FALSE;
        emvPurse.sequenceFlag[EMV.SEQF_AAC_GENERATED]      = FALSE;

        // Preparing and sending the response
        buffer[0] = EMV.TAG_PROC_OPT; // T
        buffer[1] = (byte) 0x0E;      // L

        buffer[2] = EMV.TAG_AIP; // T
        buffer[3] = (byte) 0x02; // L
        Util.arrayCopy(emvPurse.getAIP(), (short)0,
                    buffer, (short)4, EMV.AIP_LEN); // V

        buffer[6] = EMV.TAG_AFL; // T
        buffer[7] = (byte) 0x08; // L
        Util.arrayCopy(emvPurse.getAFL(), (short)0,
                    buffer, (short)8, EMV.AFL_LEN);
        apdu.setOutgoingAndSend((short)0, (short)16);
        break;

case EMV.INS_GENERATE_AC:

        len = (short)(buffer[ISO7816.OFFSET_LC] & 0x00FF);
        if(len != apdu.setIncomingAndReceive() )
    ISOException.throwIt( ISO7816.SW_WRONG_LENGTH );
        if (buffer[ISO7816.OFFSET_LC] != EMV.LC_GAC)
            ISOException.throwIt( ISO7816.SW_WRONG_LENGTH );
```

```
        // Verify P2
        if (buffer[ISO7816.OFFSET_P2] != EMV.P2_GENERATE_AC)
            ISOException.throwIt(EMV.SW_WRONG_P1P2);

        // Only TC or AAC requests are supported
        if ( (buffer[ISO7816.OFFSET_P1] != EMV.CODE_TC) &&
             (buffer[ISO7816.OFFSET_P1] != EMV.CODE_AAC) )
              ISOException.throwIt(EMV.SW_WRONG_P1P2);

        // Verify the transaction context
        if ((emvPurse.sequenceFlag[EMV.SEQF_GETPROC_PERFORMED] ==
        FALSE) || (emvPurse.sequenceFlag[EMV.SEQF_AAC_GENERATED] ==
        TRUE) )
        // ... then return "conditions of use are not satisfied"
            ISOException.throwIt(EMV.SW_COND_NOTSAT);

        // If it is the second Generate AC command...
        if ( emvPurse.sequenceFlag[EMV.SEQF_ARQC_GENERATED] == TRUE )
        // ... then return "conditions of use are not satisfied"
            ISOException.throwIt(EMV.SW_COND_NOTSAT);

        // Perform card risk management and generate respective
        // response

        response = emvPurse.processAC_I(buffer[ISO7816.OFFSET_P1],
        buffer, ISO7816.OFFSET_CDATA, buffer[ISO7816.OFFSET_LC]);

        Util.arrayCopy(response, (short)0, buffer, (short)0,
                    (short)(response[1]+2));
        // reset sequence flags
        emvPurse.sequenceFlag[EMV.SEQF_GETPROC_PERFORMED] = FALSE;

        // send the response out
        apdu.setOutgoingAndSend((short)0, (short)(response[1]+2));
    break;

    default:
        ISOException.throwIt(ISO7816.SW_INS_NOT_SUPPORTED);
    } // switch EMV INS
    break;
default:
    ISOException.throwIt(ISO7816.SW_CLA_NOT_SUPPORTED);
} // switch CLA
}
} // end of EMVdemo class
```

```
/*- - - - - - - - - - - - - - - - - - - - - - - - - - - */
/* File EMVPurse.java                                    */
/* Contains implementation of the classes:              */
/*  - EMVPurse                                           */
/*  - CVR                                                */
/*- - - - - - - - - - - - - - - - - - - - - - - - - - - */

package emvdemo;

import javacard.framework.*;
import com.gieseckedevrient.javacardx.crypto.*;
public class EMVPurse {

/*- - - - - - - - - - - - - - - - - - - - - - - - - - - */

    public final static byte TRUE  = 1;
    public final static byte FALSE = 0;

    // Definition of the application-related constants and objects

    private final static byte FILES_NUMBER = 10;
            // number of files supported
    private final static byte RECORDS_NUMBER = 15;
            // number of records in a file
    private final static byte RECORD_LENGTH = 64; // record length

    private final static byte  COUNTRY_CODE = 40;            // Austria
    private final static short CURRENCY_CODE = (short) 978; // Euro

    // Maximum allowed transaction amount (Euro)
    private final static short MAX_TRANS_AMOUNT  = (short) 650;
    // Maximum allowed transaction amount (Euro)
    private final static short MAX_CUMUL_AMOUNT = (short) 650; // in Euro

    // Application version number
    private final static byte[] AVN = { (byte) 0x00, (byte) 0x02 } ;
            // EPI ICC a.p.
    private final static byte  AVN_LEN = (byte)  0x02;

    // Application usage control
    private final static byte[] AUC = { (byte) 0xFF, (byte) 0x00 } ;
    private final static byte  AUC_LEN = (byte)  0x02;

    // Application effective date (YYMMDD)
    private final static byte[] AED = { (byte) 0x00, (byte) 0x01, (byte) 0x01} ;
    private final static byte  AED_LEN = (byte)  0x03;
```

```java
// Application expiration date (YYMMDD)
private final static byte[] AXD = { (byte) 3, (byte) 12, (byte) 31} ;
private final static byte   AXD_LEN = (byte)  0x03;


// Issuer country code
private final static byte[] ICD = { (byte) 0, COUNTRY_CODE } ;
private final static byte   ICD_LEN = (byte)  0x02;


// Application currency code
private final static byte[] ACD = { (byte) 0x03, (byte) 0xD2 } ; // Euro
private final static byte   ACD_LEN = (byte)  0x02;


// Cardholder name
private final static byte[] CHN = {'H', 'a', 'n', 's', ' ',
                                   'M', 'u', 's', 't', 'e', 'r',
                                        'm', 'a', 'n', 'n'} ;

private final static byte   CHN_LEN = (byte)  15;


// Cardholder verification method list
private final static byte[] CVM = {0x00, 0x00, 0x00, 0x00, // no X
                              0x00, 0x00, 0x00, 0x00, // no Y
                           0x41, 0x03} ;  // off-line, plaintext PIN

private final static byte  CVM_LEN = (byte) 10;



// EMV application specific values
// AIP, dynamic off-line auth, CHV
private final byte[] AIP = { (byte) 0x5C,(byte) 0x00 } ;


// AFL, two AEFs with SFIs 1 and 2
private final byte[] AFL = {   (byte) 0x08,(byte) 0x01,
                              (byte) 0x06,(byte) 0x00,
                              (byte) 0x10,(byte) 0x01,
                              (byte) 0x02,(byte) 0x00 } ;


private final byte [] DEMO_PIN = { (byte) 0x24, (byte) 0x12, (byte) 0x34,
                                  (byte) 0xFF, (byte) 0xFF, (byte) 0xFF,
                                  (byte) 0xFF, (byte) 0xFF } ;


// Card Issuer Action Code - Decline

private final byte [] CIACD = {   (byte) 0x00, (byte) 0x02, (byte) 0x40,
                                  (byte) 0x02 } ;
```

```
// master key value for the cryptogram calculation
private final byte[] MKac_VALUE =
                { (byte)0x01, (byte)0x33, (byte)0x02, (byte)0x11,
                  (byte)0x01, (byte)0x33, (byte)0x02, (byte)0x11,
                  (byte)0x11, (byte)0x33, (byte)0x02, (byte)0x01,
                  (byte)0x11, (byte)0x33, (byte)0x02, (byte)0x01 } ;

// Application elementary files descriptors
private final byte AEF1 = 1;
private final byte AEF2 = 2;


/*- - - - - - - - - - - - - - - - - - - - - - - - - - - - */




// Internal data objects
private EMVFileSystem filesystem;      // Application file system
private short ATC = (short) 0;         // Application transaction counter
private short cumulativeAmount = 0;    // Cumulative transactions amount
private OwnerPIN emvPIN;               // EMV card application PIN object
private CVR appCVR;                    // CVR object
public  byte[] sequenceFlag = null;    // Application sequence flag
private byte[] message;                // Input to AC calculation
private byte[] cryptogram;
private byte[] SKl, SKr;               // session keys L and R
private byte[] rand;          // random number for session key derivation

private SymmetricKey desKey;      // symmetric key used by DES cipher
private SymmetricKey des3Key;     // symmetric key used by DES3 cipher
private CipherECB cipherDES;      // DES cipher
private CipherECB cipherDES3;     // DES3 cipher

private byte[] record;
private byte[] GAC1_result;
private short amount;


/*- - - - - - - - - - - - - - - - - - - - - - - - - - - - */


// Class constructor
public EMVPurse () {

    // Create transient sequence flag object
    sequenceFlag = JCSystem.makeTransientByteArray(EMV.SEQF_LENGTH,
        JCSystem.CLEAR_ON_RESET);

    // Create application PIN object
    emvPIN = new OwnerPIN(EMV.PIN_TRY_LIMIT, EMV.PIN_MAX_LEN);
    emvPIN.update(DEMO_PIN, (short)0, (byte)8);
```

```
// Create cipher key objects
desKey  = new SymmetricKey((short)8, JCSystem.CLEAR_ON_RESET);
des3Key = new SymmetricKey((short)16,JCSystem.CLEAR_ON_RESET);


// Create DES and DES3 cipher objects in ECB mode
cipherDES  = CipherECB.getInstance(Cipher.ENGINE_DES);
cipherDES3 = CipherECB.getInstance(Cipher.ENGINE_3DES);


appCVR = new CVR();


message = new byte[EMV.MESSAGE_LEN];
rand = new byte[EMV.SKEY_LEN];
SKl  = new byte[EMV.SKEY_LEN];
SKr  = new byte[EMV.SKEY_LEN];
cryptogram = new byte[EMV.AC_LEN];


record = new byte[RECORD_LENGTH];
GAC1_result = new byte[EMV.GAC1_RESP_LEN];


// Creating the file system and files
filesystem = new EMVFileSystem(FILES_NUMBER);


filesystem.createFile(AEF1, (byte) 6, RECORD_LENGTH);
filesystem.createFile(AEF2, (byte) 2, RECORD_LENGTH);


// Putting EMV TLV objects into files

// Mandatory data AEF1
record = fillEMVRecord(EMV.TAG_AVN, AVN_LEN, AVN);
filesystem.writeRecord(AEF1, (byte)1, record, (byte)(record[1] + 2));


record = fillEMVRecord(EMV.TAG_AUC, AUC_LEN, AUC);
filesystem.writeRecord(AEF1, (byte)2, record, (byte)(record[1] + 2));


record = fillEMVRecord(EMV.TAG_AED, AED_LEN, AED);
filesystem.writeRecord(AEF1, (byte)3, record, (byte)(record[1] + 2));


record = fillEMVRecord(EMV.TAG_AXD, AXD_LEN, AXD);
filesystem.writeRecord(AEF1, (byte)4, record, (byte)(record[1] + 2));


record = fillEMVRecord(EMV.TAG_ICD, ICD_LEN, ICD);
filesystem.writeRecord(AEF1, (byte)5, record, (byte)(record[1] + 2));


record[1] = (byte) (CVM_LEN + 2);
record[2] = EMV.TAG_CVM;
record[3] = CVM_LEN;
Util.arrayCopy(EMVPurse.CVM, (short)0, record, (short)4, (short)CVM_LEN);
filesystem.writeRecord(AEF1,(byte)6, record, (byte)(record[1] + 2));
```

```java
        // Optional data AEF2
        record = fillEMVRecord(EMV.TAG_CHN, CHN_LEN, CHN);
        filesystem.writeRecord(AEF2, (byte)1, record, (byte)(record[1] + 2));

        record = fillEMVRecord(EMV.TAG_ACD, ACD_LEN, ACD);
        filesystem.writeRecord(AEF2, (byte)2, record, (byte)(record[1] + 2));

    }  // constructor

/*- - - - - - - - - - - - - - - - - - - - - - - - - - - - - */

// Methods

// Core method for application cryptogram generation
public byte[] processAC_I(byte request, byte[] cdata,
                          byte offset, byte len) {

        short status;
        byte action;
        byte cid;

        // Reset the CVR
        appCVR.reset();

        // Perform card risk management and card action analysis
        action = riskManagement(request, cdata, offset, len);

        // Fill T and L of the response
        GAC1_result[0] = EMV.TAG_RESP;
        GAC1_result[1] = EMV.LEN_GAC1;

        // Fill Cryptogram information data
        GAC1_result[EMV.OFFSET_CID]   = (byte) (EMV.TAG_CID >> 8);
        GAC1_result[EMV.OFFSET_CID+1] = (byte) ((EMV.TAG_CID << 8) >> 8);
        GAC1_result[EMV.OFFSET_CID+2] = EMV.LEN_CID;

        cid = action;
        if ((appCVR.getByte((byte)2) & CIACD[2]) == CIACD[2]) // PIN try
                 limit exceeded
            cid = (byte) (cid | 2); // set bit 2

        GAC1_result[EMV.OFFSET_CID+3] = cid;

        // fill ATC
        GAC1_result[EMV.OFFSET_ATC]   = (byte) (EMV.TAG_ATC >> 8);
        GAC1_result[EMV.OFFSET_ATC+1] = (byte) ((EMV.TAG_ATC << 8) >> 8);
        GAC1_result[EMV.OFFSET_ATC+2] = EMV.LEN_ATC;
        GAC1_result[EMV.OFFSET_ATC+3] = (byte) (ATC >> 8);
        GAC1_result[EMV.OFFSET_ATC+4] = (byte) ((ATC << 8) >> 8);
```

```
// prepare the cryptogram message
Util.arrayCopy(cdata, ISO7816.OFFSET_CDATA, message, (short)0,
    (short)29);
Util.arrayCopy(AIP, (short)0, message, (short)29, (short)2);
Util.setShort(message, (short)31, ATC);
Util.arrayCopy(appCVR.getBytes(), (short)0, message, (short)33,
    (short)4);


// pad the message according to ISO/IEC 9797, method 2
message[37] = (byte)0x80; message[38] = (byte)0;
message[39] = (byte)0;


// derive session keys
// first, prepare the input for the key derivation function
Util.setShort(rand,(short)0,ATC);
rand[2] = (byte)0; rand[3] = (byte)0;
Util.arrayCopy(cdata, (short)(ISO7816.OFFSET_CDATA+EMV.OFFSET_TUN),
                rand, (short)4,(short)4);


des3Key.setValue(MKac_VALUE,(short)0,(short)16);
cipherDES3.setKey(des3Key);
derive_SKl(cipherDES3, rand, SKl);
derive_SKr(cipherDES3, rand, SKr);


// compute cryptogram
compute_ac(message,EMV.MESSAGE_LEN,SKl,SKr,cipherDES,desKey,cryptogram);


// fill cryptogram in
GAC1_result[EMV.OFFSET_AC]   = (byte) (EMV.TAG_AC >> 8);
GAC1_result[EMV.OFFSET_AC+1] = (byte) ((EMV.TAG_AC << 8) >> 8);
GAC1_result[EMV.OFFSET_AC+2] = EMV.LEN_AC;
Util.arrayCopy(cryptogram,(short)0,GAC1_result,
                (short)(EMV.OFFSET_AC+3),(short)EMV.AC_LEN);


// fill Issuer Application data
GAC1_result[EMV.OFFSET_IAD]   = (byte) (EMV.TAG_IAD >> 8);
GAC1_result[EMV.OFFSET_IAD+1] = (byte) ((EMV.TAG_IAD << 8) >> 8);
GAC1_result[EMV.OFFSET_IAD+2] = EMV.LEN_IAD;
GAC1_result[EMV.OFFSET_IAD+3] = (byte) 1; // key derivation index
GAC1_result[EMV.OFFSET_IAD+4] = (byte) 1; // cryptogram version number

Util.arrayCopy(appCVR.getBytes(), (short)0, GAC1_result,
                (short)(EMV.OFFSET_IAD+5),(short)4);


return GAC1_result;

}  // process AC
```

```
// Derivation of the left part of the session key
private void derive_SKl(CipherECB cipher, byte[] input, byte[] output) {

    input[2] = (byte)0xF0;
    cipher.encrypt(input, (short)0, (short)8, output, (short)0,
                    CipherECB.PADDING_ISO00);
    return;

}  // derive_SKl


// Derivation of the right part of the session key
private void derive_SKr(CipherECB cipher, byte[] input, byte[] output) {

    input[2] = (byte)0x0F;
    cipher.encrypt(input, (short)0, (short)8, output, (short)0,
                    CipherECB.PADDING_ISO00);
    return;


}  // derive_SKr

// Computing the application cryptogram, return value is ac byte array
private void compute_ac(byte[] message, byte mesLen, byte[] keyL, byte[] keyR,
                    CipherECB cipher, SymmetricKey key, byte[] ac) {

    byte rounds, j, i;

    rounds = (byte)(mesLen / 8);
    key.setValue(keyL, (short)0, (short)8);
    cipher.setKey(key);

    for (i=0; i < 8; i++)
        ac[i]=message[i];
    for (j=1; j < (byte)(rounds+1); j++) {
        cipher.encrypt(ac,(short)(0), (short)8, ac, (short)0,
            CipherECB.PADDING_ISO00);
        if (j != rounds)
            for (i=0; i < 8; i++)
                ac[i] = (byte)(ac[i]^message[(byte)(j*8+i)]);
    }
    key.setValue(keyR, (short)0, (short)8);
    cipher.setKey(key);
    cipher.decrypt(ac,(short)0, (short)8, ac, (short)0);

    key.setValue(keyL, (short)0, (short)8);
    cipher.setKey(key);
    cipher.encrypt(ac,(short)(0), (short)8, ac, (short)0,
            CipherECB.PADDING_ISO00);
```

```
        return;
}   // compute_ac


// Card risk management routine
private byte riskManagement(byte request, byte[] cdata, byte offset, byte len) {

    byte action;
    byte[] cvrBytes;

    // if ACC was requested, the card answers with AAC
    // or if any other from TC was requested - answer with AAC
    if ( (request == EMV.CODE_AAC) || (request != EMV.CODE_TC) ) {
        // set relevant CVR bits
        appCVR.setGAC2notReq();
        appCVR.setAACinGAC1();
        return EMV.CODE_AAC;
    }
    // Card Risk Management functions
    // PIN verification status function
    if (sequenceFlag[EMV.SEQF_PIN_PERFORMED] == TRUE)
        appCVR.setPINPerformed(); // VERIFY command was given during the
        transaction

    if ( (sequenceFlag[EMV.SEQF_PIN_VERIFIED] == FALSE) &&
        (sequenceFlag[EMV.SEQF_PIN_PERFORMED] == TRUE))
        appCVR.setPINFailed(); // PIN verification failed

    if (emvPIN.getTriesRemaining() == 0)
        appCVR.setPINTryLimit(); // PIN Try Limit exceeded


    // Maximum Offline transaction amount check function
    amount = Util.makeShort(cdata[offset+4],cdata[offset+5]);
    // if the amount exceeds the maximum
    if ((Util.makeShort(cdata[offset+19],cdata[offset+20]) == CURRENCY_
        CODE) && (amount  MAX_TRANS_AMOUNT))
        appCVR.setMaxAmount();

// check that higher digits of the Amount field are 00
    if ( (cdata[offset] != 0) ||
        cdata[offset+1] !=0) || (cdata[offset+2] !=0) ||
            (cdata[offset+3] !=0) )
        appCVR.setMaxAmount();


    // Maximum cumulative amount check function
    // (if native-currency transaction)
    if ((Util.makeShort(cdata[offset+19],cdata[offset+20]) == CURRENCY_
            CODE) && (amount + cumulativeAmount  MAX_CUMUL_AMOUNT))
        appCVR.setMaxAmount();
```

```
// Perform card action analysis


// verify the CVR against CIAC - decline
cvrBytes = appCVR.getBytes();
if ( ((cvrBytes[1] & CIACD[1]) == CIACD[1]) || // PIN Verification failed
((cvrBytes[2] & CIACD[2]) == CIACD[2]) || // PIN Try limit exceeded
((cvrBytes[3] & CIACD[3]) == CIACD[3]) || // Upper consecutive or
                                    // Accumulative amount exceeded
    (sequenceFlag[EMV.SEQF_PIN_PERFORMED] == FALSE)) // No PIN Verify
{
    // set relevant CVR bits
        appCVR.setGAC2notReq();
        appCVR.setAACinGAC1();


    action = EMV.CODE_AAC; // -- decline transaction
}
else
{
if (Util.makeShort(cdata[offset+19],cdata[offset+20]) == CURRENCY_CODE)
        cumulativeAmount = (short)(cumulativeAmount + amount);

    // set relevant CVR bits
    appCVR.setGAC2notReq();
    appCVR.setTCinGAC1();


    action = EMV.CODE_TC; // — complete transaction (issue TC)
}

return action;
}  // card risk management


// Method fillEMVRecord: fills a TLV object to a binary array
private byte[] fillEMVRecord(short t, byte l, byte[] v)
{
    byte[] rec = new byte[EMVPurse.RECORD_LENGTH];

    rec[0] = EMV.TAG_RECORD;
    rec[1] = (byte) (l + 3);
    rec[2] = (byte) (t >> 8);
    rec[3] = (byte) ((t << 8) >> 8);
    rec[4] = l;

    Util.arrayCopy(v, (short)0, rec, (short)5, (short)l);

    return rec;

}  // fillENVRecord
```

```
    // Method readRecord: reads a record from the given AEF
    public EMVFileRecord readRecord(byte sfi, byte recnum) {
        return filesystem.readRecord(sfi,recnum);
    }


    // Method getPINTriesRemaining: returns the application PIN
    // try counter
    public byte getPINTriesRemaining() {
        return emvPIN.getTriesRemaining();
    }


    // Method checkPIN: verifies the application PIN against a given value
    public boolean checkPIN(byte[] buffer, short offset, byte len) {
        return emvPIN.check(buffer, offset, len);
    }


    // Method getATC: returns the current value of ATC
    public short getATC() {
        return ATC;
    }


    // Method incrementATC: increments ATC value by 1
    public void incrementATC() {
        ++ATC;
    }


    // Method getAFL: returns AFL value
    public byte[] getAFL() {
        return AFL;
    }


    // Method getAIP: returns AIP value
    public byte[] getAIP() {
        return AIP;
    }

} // class EMVPurse

// class representing the CVR (Card Verification Results object)

class CVR {
    private byte[] bytes;
    // constructor
    public CVR () {
        bytes = new byte[4];
        bytes[0] = 3; // CVR length is given in the first byte
    }
```

```
// method getBytes, returns CVR bytes
public byte[] getBytes() {
    return bytes;
}
// method getByte, returns a particular byte of CVR
public byte getByte (byte n) {
    return bytes[n];
}
// method reset, resets the CVR content
public void reset() {
    bytes[1] = 0;
    bytes[2] = 0;
    bytes[3] = 0;
}


// method setPINPerformed(), sets the "Offline PIN verification was performed"
public void setPINPerformed() {
    // set bit 3 in byte 2
    bytes[1] = (byte)(bytes[1] | 4);
}


// method setPINFailed(), sets the "Offline PIN verification failed"
public void setPINFailed() {
    // set bit 2 in byte 2
    bytes[1] = (byte)(bytes[1] | 2);
}


// method setPINTryLimit(), sets the "PIN Try Limit exceeded"
public void setPINTryLimit() {
    // set bit 7 in byte 3
    bytes[2] = (byte)(bytes[2] | 64);
}


// method setMaxAmount(), sets the "Maximum transaction amount exceeded"
public void setMaxAmount() {
    // set bit 2 byte 4
    bytes[3] = (byte)(bytes[3] | 2);
}


// method setGACnotReq(), sets the "Second Generate AC not requested"
public void setGAC2notReq() {
    // set bit 8 in byte 2
    bytes[1] = (byte)(bytes[1] | 128);
}


// method setTCinGAC1(), sets the "TC returned in the first Generate AC
public void setTCinGAC1() {
    // set bit 5 in byte 2
    bytes[1] = (byte)(bytes[1] | 16);
}
```

```
    // method setAACinGAC1(), sets the "AAC returned in the first Generate AC
    public void setAACinGAC1() {
        // nothing to set (relevant bits are 0)
        return;
    }

}  // class CVR
```

```
/*- - - - - - - - - - - - - - - - - - - - - - - - - - - - */
/* File EMVdemo.java                                       */
/* Contains implementation of the                         */
/* on-card file system classes                            */
/*                                                         */
/*- - - - - - - - - - - - - - - - - - - - - - - - - - - - */
package emvdemo;

import javacard.framework.*;


// Implementation of the EMV card file system
// Class EMVfile defines a linear fixed-record file for storing
// EMV application data objects


class EMVFileSystem
{
    private byte[] selected_flag;
    private EMVFile[] files;
    private byte files_num;   // Maximum number of files supported
    private byte next_av = 0; // next file that can be created
    private byte selected_sfi; // SFI of the currently selected file

    public EMVFileSystem (byte maxfiles)
    {
        files_num = maxfiles;
        files = new EMVFile[maxfiles];
        selected_flag =
            JCSystem.makeTransientByteArray((short)1,JCSystem.CLEAR_ON_RESET);
    }


    public void createFile (byte sfi, byte recnum, byte reclen) throws ISOException
    {
        if (next_av == files_num)
            ISOException.throwIt(ISO7816.SW_FILE_INVALID);

        files[next_av] = new EMVFile(sfi, recnum, reclen);
        next_av++;
    }



    public boolean selectFile(byte sfi)
    {
        byte i;

        for (i=0; i next_av; i++)
            if (files[i].getSFI() == sfi)
            {
                selected_flag[0] = (byte) 0xFF;
                selected_sfi = sfi;
```

```
                        return true;
                }
            return false;
    }


    public EMVFileRecord readRecord (byte sfi, byte recnum) throws ISOException
    {
        byte i;
        boolean f = false;
        for (i=0; i < next_av; i++)
            if (files[i].getSFI() == sfi)
            {
                f = true;
                break;
            }

        if (!f)
            ISOException.throwIt(ISO7816.SW_FILE_NOT_FOUND);

return files[i].readRecord((byte)(recnum - 1)); // EMV rec num starts with 1!!
    }


    public void writeRecord (byte sfi, byte recnum, byte[] value, byte len)
        throws ISOException
    {
        byte i;

        for (i=0; i next_av; i++)
            if (files[i].getSFI() == sfi)
            {
            files[i].writeRecord((byte)(recnum - 1), value, len); // EMV rec num
                starts with 1!! return;
            }

            ISOException.throwIt(ISO7816.SW_FILE_NOT_FOUND);
    }


}  // Class EMVFileSystem


class EMVFile
{

    private byte rec_number; // number of records in the file
    private EMVFileRecord[] file_records; // records themselves
    private byte _SFI;

    public EMVFile (byte sfi, byte rn, byte rl) throws ISOException
    {
        byte i;
```

```
        if (rn  EMV.MAX_FILE_SIZE)
            ISOException.throwIt(EMV.SW_MEM_FAILURE);

        _SFI = sfi;

        file_records = new EMVFileRecord[rn];

        for (i=0; i < rn; i++)
            file_records[i] = new EMVFileRecord(rl);

        rec_number = rn;
    }

    public byte getSFI()
    {
        return _SFI;
    }

    // method for reading a record
    public EMVFileRecord readRecord (byte recnum) throws ISOException
    {
        if (recnum > rec_number-1)
            ISOException.throwIt(ISO7816.SW_RECORD_NOT_FOUND);

            return file_records[recnum];
    }

    // method for writing data to existing record
    public void writeRecord (byte recnum, byte[] value, byte len)
        throws ISOException
    {
        if (recnum > rec_number - 1)
            ISOException.throwIt(ISO7816.SW_RECORD_NOT_FOUND);

        if (len > file_records[recnum].getRecordLen() - 1)
            ISOException.throwIt(ISO7816.SW_WRONG_LENGTH);

        file_records[recnum].writeData(value, len);
    }

    public byte getRecordsNum()
    {
        return rec_number;
    }
}


class EMVFileRecord
{
```

```
private byte rec_length; // maximum length of the record
private byte act_length; // actual length of the data in the record

private byte[] record_data;


public EMVFileRecord (byte rl)
{
    record_data = new byte[rl];
    rec_length = rl;
}

public void writeData (byte[] value, byte len)
{
    JCSystem.beginTransaction();
    Util.arrayCopy(value, (short)0, record_data, (short)0, (short)len);
    act_length = len;
    JCSystem.commitTransaction();
}

public byte getRecordLen()
{
    return rec_length;
}

public byte getActualLen()
{
    return act_length;
}

public byte[] getData()
{
    return record_data;
}
}
```

Appendix B:
OCF Reference Manual

This is intended as a reference manual to the most important OCF classes. There are too many OCF classes to describe them all here, so only classes that are important to the developer or are the backbone of OCF structure are mentioned here [1]. From each of the classes, the public methods/fields are described. Private methods/fields and deprecated methods/fields are not covered.

B.1 Package `opencard.core.service`

B.1.1 Class `SmartCard`

```
public final class SmartCard
extends java.lang.Object
```

The `SmartCard` object is the point of access to OCF for the application. It provides the card service for the application and has an associated card service scheduler that manages communication with the card terminal.

Constructor

```
public SmartCard(CardServiceScheduler scheduler,
CardID cid)
```

Makes a `SmartCard` object and associates a `CardServiceScheduler` object to it.

Parameters

- `scheduler`—the `CardServiceScheduler` that controls the object;
- `cid`—the `CardID` object representing the `Smart Card` itself.

Methods

```
public static java.lang.String getVersion()
```

Returns the OCF version information.

```
public void beginMutex()
      throws java.lang.InterruptedException,
        CardTerminalException
```

Begins a section where you will have exclusive access to the `SmartCard`. Somewhere at the end, `endMutex` has to be called. The `close` method will do this anyway.

```
public void close()
        throws CardTerminalException
```

Closes this `SmartCard` object, cleaning up everything that might be associated with the object.

```
public void endMutex()
```

Signals the end of exclusive access.

```
public CardID getCardID()
```

Returns the `CardID`. Useful for obtaining ATR and so on.

```
public CardService getCardService (java.lang. Class
                                    clazz, boolean block)
      throws java.lang.ClassNotFoundException,
        CardServiceException
```

This method tries to generate a card service of the type `clazz`, from which an instance is returned.

Parameters

- `clazz`—the class that the `CardService` shall implement;
- `block`—decides if running in blocking mode.

```
public static SmartCard
    getSmartCard(CardTerminalEvent ctEvent,
                 CardRequest req)
        throws CardTerminalException
```

This will generate a `SmartCard` object when a certain event (card insertion) occurs. Used for event-driven OCF programs.

Parameters

- `ctEvent`—the received `CardTerminalEvent`;
- `req`—a `CardRequest` object to see whether the card that was inserted can be handled by our application.

```
public static SmartCard
    getSmartCard(CardTerminalEvent ctEvent,
                 CardRequest req,
                 java.lang.Object lockHandle)
        throws CardTerminalException
```

Same as above but with an additional parameter that provides a handle obtained by the lock owner when locking a slot or terminal.

```
public static boolean isStarted()
```

Indicates whether the global start-up process has been finished.

```
public static void shutdown()
    throws CardTerminalException
```

Global method to shut down OCF entirely. Must be done if the application exits, whether or not it was successful. It releases the serial port and cleans up the registries.

```
public static void start()
    throws OpenCardPropertyLoadingException,
    java.lang.ClassNotFoundException,
    CardServiceException,
    CardTerminalException
```

Global method to initialize OCF. Must be invoked before any OCF-related steps can be taken. Initializes the registries in particular.

```
public static SmartCard waitForCard(CardRequest req)
    throws CardTerminalException
```

Waits for a card to be inserted. If the card request can be fulfilled, a Smart-Card object is returned, otherwise NULL. Unlike the event-driven methods, this method will block until a card is inserted (or an exception occurs).

```
public static SmartCard waitForCard(CardRequest req,
    java.lang.Object lockHandle)
        throws CardTerminalException
```

Like above, but for locked terminals.

B.1.2 Class CardRequest

```
public class CardRequest
extends java.lang.Object
```

A CardRequest determines whether the card that is inserted can be handled by the application; it is therefore essential to the application flow.

Fields

```
public static final int ANYCARD
```

Wait behavior if a card is already inserted or waiting for a new card. Default.

```
public static final int NEWCARD
```

A new card must be inserted. Any already inserted cards are not recognized.

Constructor

```
public CardRequest(int waitBehavior,
                   CardTerminal terminal,
                   java.lang.Class cardServiceClass)
```

Creates a new CardRequest.

Parameters

- waitBehavior—one of the two constants specified in the fields section;
- terminal—specify in which terminal the card has to be inserted, or NULL for any terminal;
- cardServiceClass—specify which card services must be supported by the card, or NULL for "don't care at this time."

Methods

```
public void setTimeout(int timeout)
```

You can set a timeout value (in seconds) to wait for a card. If the timeout value is reached, waitForCard of SmartCard object returns NULL. A value of −1 will unset the timeout.

```
public void setFilter(CardIDFilter filter)
```

Sets the filter of the CardRequest that a card must get through. Seldom used.

```
public CardIDFilter getFilter()
```

Gets the filter that was set in the previous method.

```
public java.lang.Class getCardServiceClass()
```

Gets an object representing the card service.

```
public CardTerminal getCardTerminal()
```

Gets the CardTerminal object.

```
public int getTimeout()
```

Gets the timeout value.

```
public int getWaitBehavior()
```

Returns the wait behavior.

```
public boolean isTimeoutSet()
```

Determines whether a timeout period is set.

```
public java.lang.String toString()
```

Returns a string representation of this card request.

B.1.3 Class `CardServiceFactory`

```
public abstract class CardServiceFactory
extends java.lang.Object
```

A `CardServiceFactory` produces card services for a specific smart card. Typically, the `CardServiceRegistry` will instantiate a `CardService-Factory` once a smart card has been inserted for which a `waitForCard` method was invoked. Subclasses must implement the `getCardType` and `getClasses` methods, which can communicate with the card to classify the card.

Card service factories programmed in OCF 1.1 style must instead derive from `opencard.opt.service.OCF11CardServiceFactory` which offers the `knows` and `cardServiceClasses` methods that cannot communicate with the card.

Constructor

```
public CardServiceFactory()
```

Instantiates a `CardServiceFactory`.

Methods

```
protected CardService
getCardServiceInstance(java.lang.Class clazz,
```

```
                    CardType type,
                    CardServiceScheduler scheduler,
                    SmartCard card,
                    boolean block)
            throws CardServiceException
```

Instantiates a `CardService` implementing the `clazz` class.

Parameters

- `clazz`—the class object for the desired class;
- `cid`—a `CardID` object representing the actual smart card inserted;
- `scheduler`—the controlling `CardServiceScheduler`;
- `card`—the `SmartCard` object requesting the `CardService`;
- `block`—specifies the waiting behavior of the newly created `CardService`.

```
protected java.lang.Class
        getClassFor(java.lang.Class clazz,
                    CardType type)
```

Locates the `CardService` class that implements `clazz`.

Parameters

- `clazz`—the class object for the desired class;
- `type`—a `CardType` object representing the smart card for which the `CardService` is requested.

```
protected CardService
newCardServiceInstance(java.lang.Class clazz,
                    CardType type,
                    CardServiceScheduler scheduler,
                    SmartCard card,
                    boolean blocking)
            throws CardServiceException
```

Normally only used internally.

```
protected abstract CardType
    getCardType(CardID cid,
                CardServiceScheduler scheduler)
        throws CardTerminalException
```

This method must be overwritten and indicates whether the factory is able to instantiate service objects for this type of card.

Parameters

- `cid`—a `CardID` representing an ATR;

- `scheduler`—a `CardServiceScheduler` that can be used to communicate with the card to determine, for example, if a specific application is running (the information provided by the cid may be insufficient).

Returns a valid `CardType` if the factory can instantiate services for this card or `CardType.UNSUPPORTED` if the factory does not know the card.

```
protected abstract java.util.Enumeration
    getClasses(CardType type)
```

Another method that has to be overwritten. It should simply return an enumeration of all services the factory is able to produce.

Parameter

- `type`—the `CardType` of the smart card for which the enumeration is requested.

B.1.4 Class `CardServiceRegistry`

```
public final class CardServiceRegistry
extends java.lang.Object
```

The registry knows all its registered factories and picks up that which is able to produce the service object. There is only one instance of `CardServiceRegistry` in a running application.

Methods

```
public void add(CardServiceFactory factory)
```

Adds a `CardServiceFactory` to the registry.

```
protected java.lang.Class
    getCardServiceClassFor(java.lang.Class clazz,
                           CardID cid,
                           CardServiceScheduler
                           scheduler)
```

Tries to instantiate an object for the desired class. This can only succeed if there is at least one factory that knows the class.

Parameters

- `clazz`—the class that the requested `CardService` should be an instance of;
- `cid`—a `CardID` object representing the ATR for which the `Card-Service` is requested;
- `scheduler`—a `CardServiceScheduler` for the card to be inspected. This will be given to the factory so it can communicate with the card to obtain additional information.

Returns a `CardService` class object or NULL if all factories deny knowledge of the service.

```
public final java.util.Enumeration
    getCardServiceFactories()
```

Gets all registered card service factories and returns an enumeration of them.

```
protected CardService
    getCardServiceInstance(java.lang.Class clazz,
                           CardID cid,
                           CardServiceScheduler
                           scheduler,
                           SmartCard card,
                           boolean block)
        throws java.lang.ClassNotFoundException
```

Like `getCardServiceClassFor`, this method tries to instantiate a `CardService`, but here the real object is returned instead of its class representation.

Parameters

- `clazz`—the class that the requested `CardService` should be an instance of;
- `cid`—a `CardID` object representing the ATR for which the `Card-Service` is requested;
- `scheduler`—a `CardServiceScheduler` for the card to be inspected. This will be given to the factory so it can communicate with the card to obtain additional information;
- `card`—the `SmartCard` object requesting the `CardService`;
- `block`—specifies the waiting behavior of the newly created `Card-Service`.

```
public static CardServiceRegistry getRegistry()
```

Gets a reference to the system-wide `CardServiceRegistry` object.

```
protected SmartCard
    getSmartCard(CardTerminalEvent ctEvent,
                 CardRequest req,
                 java.lang.Object lockHandle)
        throws CardTerminalException
```

Tries to get a `SmartCard` object based on a received `CardTerminal-Event`.

Parameters

- `ctEvent`—a `CardTerminalEvent` event received from a terminal;
- `req`—a `CardRequest` object describing what kind of `SmartCard` is requested;
- `lockHandle`—a handle obtained by lock owner when locking the terminal.

```
public void remove(CardServiceFactory factory)
```

Removes the passed `CardServiceFactory` from the registry.

```
public java.lang.String toString()
```

Gets a string representation of this `CardServiceRegistry`.

B.1.5 Class `CardService`

```
public abstract class CardService
extends java.lang.Object
```

An abstract prototype for a card service from which every card service that is implemented has to derive. It implements common, basic functionality. Any `SmartCard` commands (SELECT, READ RECORD, etc.) must be implemented by the card service developer in a concrete service class which is specific to the card operating system. Communication with a smart card takes place through a `CardChannel` that the card service either allocates from a `CardServiceScheduler` or gets from a third party (e.g., another card service or the corresponding `SmartCard` object if `beginMutex` is invoked there).

The methods to allocate and release card channels provided here are aware of channels that have been preset by a third party. A public method providing card functionality in a class derived from `CardService` will typically have the following structure (pseudocode):

```
public ReturnType doSomeThingWithCard(...)
    throws CardServiceException, CardTerminalException
{
    Check_Parameters();

    allocateCardChannel(); // if not allocated by
    initialization
    CommandAPDU command = ...; // has to be built.
    ResponseAPDU response =
        getCardChannel().sendCommandAPDU(command);
    Evaluate_Response();
    releaseCardChannel(); // free if allocated in this
    routine, else skip this step
    return retvalue;
}
```

Constructor

```
protected CardService()
```

Creates a new card service. Before you can use it, initialize has to be invoked.

Methods

```
public void setCardChannel(CardChannel channel)
```

Sets the channel to use for communicating with the smart card. Setting the channel with this method avoids allocating a channel before each operation and releasing it afterward. This can be used to issue a series of commands to the smart card that has to or should be processed without intervening commands.

This method is typically invoked from SmartCard.beginMutex. All services used by an application will be provided with the same channel, and no other application will have access to the smart card.

Presetting a channel does not mean that allocateCardChannel and releaseCardChannel cannot be invoked. Their implementation in this class merely avoids the invocations of the scheduler if a channel is already available. They still check whether the scheduler is alive, or whether it has died since the smart card has been removed.

Parameter

- channel—the channel to use, or NULL to reset.

```
public final CardChannel getCardChannel()
```

Gets the card channel to use for communicating with the smart card. Returns the channel for communication with the smart card.

```
public void setCHVDialog(CHVDialog dialog)
```

Sets the CHV dialog to be used for getting passwords (PIN) from the user.

Parameter

dialog—the CHV dialog to be used.

```
public final CHVDialog getCHVDialog()
```

Returns the dialog object for CHV input.

```
public final SmartCard getCard()
```

Gets the SmartCard object associated with this service. Services are requested at a particular instance of SmartCard, which can be used to identify the smart card and also represents the application that requested the service. This method returns the SmartCard object that was used to request this service.

```
protected void initialise(CardServiceScheduler
                          scheduler,
                          SmartCard smartcard,
                          boolean blocking)
        throws CardServiceException
```

Initializes this service. This method is an extension to the constructor. It is invoked by the CardServiceFactory after creation of a service using the default constructor. The service cannot be used until this method has been invoked and returned without throwing an exception.

Derived services may override this method to perform extended initialization (e.g., allocate a card channel). However, the implementation in this class has to be invoked in any case. The preferred way to do this is by invoking super.initialize at the beginning of the redefined method.

Parameters

- scheduler—where this service is going to allocate channels;

- smartcard—which smart card has to be supported by this service;

- blocking—whether channel allocation is going to be blocking.

```
protected void allocateCardChannel()
        throws InvalidCardChannelException
```

Allocates a card channel if one is required. If a channel has been provided by setCardChannel, none has to be allocated. The method releaseCardChannel will release the channel only if it has actually been allocated here. After this method is called, a card channel will be available and can be obtained via getCardChannel.

```
protected void releaseCardChannel()
    throws InvalidCardChannelException
```

Releases the allocated card channel. If the channel was not allocated by allocateCardChannel, but was instead provided via setCardChannel, it will not be released.

B.1.6   Class CardServiceScheduler

```
public final class CardServiceScheduler
extends java.lang.Object
implements CTListener
```

This is a manager for logical channels to an inserted smart card. For each smart card known to the system and in use by some application, there is one CardServiceScheduler that manages the access to the physical smart card.

Constructor

```
public CardServiceScheduler(SlotChannel slotchannel)
```

Instantiates a new scheduler that is tied to the given slot channel.

Methods

```
public final SlotChannel getSlotChannel()
```

Returns the slot channel for this scheduler.

```
public CardChannel
    allocateCardChannel(java.lang.Object applicant,
                        boolean block)
        throws CardTerminalException
```

Allocates a card channel.

Parameters

- applicant—the object requesting the card channel. This parameter will be useful when support for multichannel cards is implemented.

- block—If true, the calling thread will be suspended until a CardChannel becomes available; if false, allocateCard-Channel() will return NULL if no channel is available.

Returns the allocated card channel, or NULL if none has been allocated.

```
public void releaseCardChannel(CardChannel channel)
throws InvalidCardChannelException
```

Releases a card channel. The channel to release must have been allocated using allocateCardChannel.

```
public final CardID reset(CardChannel ch,
                          boolean block)
       throws CardTerminalException
```

Physically resets the card associated with this CardServiceScheduler and returns a CardID object containing the ATR of the card.

Parameters

- channel—If the caller already has a channel he can provide it; otherwise the scheduler will allocate the channel itself.

- block—If true, the calling thread will be suspended until a CardChannel becomes available.

```
public void cardInserted(CardTerminalEvent ctEvent)
```

Dummy method. Since the scheduler is interested only in the removal of the associated smart card, card insertion events are ignored. This method has to be implemented in any case, since it is required by CTListener.

```
public void cardRemoved(CardTerminalEvent ctEvent)
    throws CardTerminalException
```

Signals that a smart card has been removed. If the removed card is the one associated with this scheduler, the scheduler will shut down. Any channels to the smart card will be closed. This method is required by the interface CTListener.

```
public java.lang.String toString()
```

Returns a string representation of this scheduler.

```
protected final boolean isAlive()
    throws CardTerminalException
```

Checks whether this scheduler is alive.

```
protected SmartCard createSmartCard(CardID cid)
    throws CardTerminalException
```

Creates a new SmartCard object. An associated reference counter is incremented.

Parameter

- cid—the CardID representing the smart card.

```
protected void releaseSmartCard(SmartCard card)
    throws CardTerminalException
```

Releases a SmartCard object. If the associated reference counter reaches 0, the SlotChannel is closed and the CardServiceRegistry informed of this fact.

Parameter

- card—the SmartCard object to release.

B.1.7 Class CardChannel

```
public class CardChannel
extends java.lang.Object
```

A communication channel to a smart card. It is used to exchange APDUs with an associated smart card. Specifically, the CardChannel is associated with a CardScheduler that manages the communication between the card service and the underlying terminal-related classes. It can be used to access the terminal into which it is inserted. CardChannel is the card service layer associated with the terminal layer's SlotChannel.

Constructor

```
protected CardChannel(SlotChannel slotchannel)
```

Instantiates a new logical card channel. The new channel has the given under-lying physical slot channel which will be used to contact the smart card.

Methods

```
public final boolean isOpen()
```

Checks whether this channel is currently open.

```
public CardTerminal getCardTerminal()
```

Returns the card terminal associated with this channel.

```
public final void setState(java.lang.Object state)
```

Stores a service-specific object associated with this channel. The stored object can be retrieved by `getState`. This method is invoked only by a card serv-ice to which this channel is currently allocated. The `state` object can, for example, be used to store the current selection.

Parameter

- `state`—the object to associate with this channel, or NULL to reset previous associations.

```
public final java.lang.Object getState()
```

Retrieves the service-specific object associated with this channel.

```
public ResponseAPDU sendCommandAPDU(CommandAPDU
cmdAPDU)
        throws InvalidCardChannelException,
            CardTerminalException
```

For the card service communicating with the card, this is probably the most important command. It is simply the challenge-response mechanism on which smart card communication is built. It has a `CommandAPDU` parameter for input and returns the appropriate `ResponseAPDU`.

```
public final ResponseAPDU
    sendVerifiedAPDU(CommandAPDU command,
                     CHVControl control,
                     CHVDialog dialog)
        throws InvalidCardChannelException,
               CardTerminalException,
               CardServiceInvalidCredentialException
```

Before sending a command to the card, this command will send a VERIFY command. The data relevant to verification has to be entered by some CHVDialog object (NULL for the default dialog set in object CardService). After successful verification, the APDU described in the command is sent. This method supports GUI password dialogs as well as a card terminal's encrypting PIN pad, if available.

The decision regarding whether the terminal or a GUI dialog is used for querying the password is crucial to security. Currently, the GUI is used only if the terminal does not implement the interface VerifiedAPDUInterface. This behavior must not be changed in derived classes, therefore this method is final.

Parameters

- command—the CommandAPDU to send;

- control—the verification parameters to use;

- dialog—the dialog object that prompts the user to enter the PIN. Ignored if the terminal takes responsibility for querying the PIN.

Returns the response from the smart card for the command specified by the parameter.

```
public void open()
    throws InvalidCardChannelException
```

Opens this CardChannel. A CardChannel can be reopened using this method unless closeFinal has been called.

```
public void close()
```

Closes this CardChannel.

```
protected void closeFinal()
```

Closes this CardChannel so that it cannot be reopened again.

```
public void finalize()
```

Tries to clean up.

```
public java.lang.String toString()
```

Returns a humanly readable representation of this CardChannel object.

B.1.8 Class CardType

```
public class CardType
extends java.lang.Object
```

The CardType represents a classification of the card according to the card classification scheme used by the card service factory. In most cases the card service factory will simply use a numeric value to classify a card. Optionally, any information describing the card type can be attached using setCardInfo.

Fields

```
public static CardType UNSUPPORTED
```

Indicates that the factory simply does not know this card and is therefore unable to produce a card service object that can communicate with it.

Constructors

```
public CardType()
```

Default constructor.

```
public CardType(int type)
```

Constructor from integer describing the type of card. Should be used by CardServiceFactory subclasses.

Methods

```
public int getType()
```

Returns the numeric type.

```
public void setInfo(java.lang.Object cardInfo)
```

Attaches additional information with the CardType that can be used when instantiating card services.

```
public java.lang.Object getInfo()
```

Returns the object set above.

B.1.9 Class DefaultCHVDialog

```
public class DefaultCHVDialog
    extends java.lang.Object
    implements CHVDialog
```

The default dialog used by card services. It will pop up a dialog window that queries for an input (echoed in asterisks), which is returned as String. A conversion to fit the concrete APDU data format has to be done by the calling method. It has an inner class that does the window handling using AWT.

Method

```
public java.lang.String getCHV(int chvNumber)
```

Pops up the window querying the PIN and returns the string representation of the result.

B.2 Package opencard.core.terminal

B.2.1 Class APDU

```
public abstract class APDU
extends java.lang.Object
```

An APDU represents an application protocol data unit, which is the basic unit of communication with a smart card. There are two subclasses of APDUs, CommandAPDU and ResponseAPDU, which are used for smart card commands and responses, respectively, but do not differ very much in functionality. The only extensions are additional functions of ResponseAPDU to read out separately the status word and the data [sw(), sw1(), sw2() and data()]

Fields

```
protected byte[] apdu_buffer
```

Here the APDU data resides.

```
protected int apdu_length
```

The actual length of the command APDU in the buffer. This may vary from 0 to the total length of the APDU.

Constructors

```
public APDU(byte[] buffer)
```

Creates a new APDU object and initializes it with the given buffer. The internal buffer's length is set to the length of the buffer passed.

```
public APDU(byte[] buffer,
            int length)
```

As above, but the internal buffer's length is set to the parameter length.

```
public APDU(int size)
```

Methods

```
public void append(byte[] bytes)
    throws java.lang.IndexOutOfBoundsException
```

Appends the given byte array to the APDU under construction.

```
public void append(byte b)
    throws java.lang.IndexOutOfBoundsException
```

As above, but adds just a single byte.

```
public final byte[] getBuffer()
```

Returns the APDU buffer. If the length of the APDU is changed by some operation (e.g., calculating and adding a MAC), setLength has to be used to store the new length.

```
public final int getByte(int index)
```

Gets the byte at the specified position in the buffer. The byte is converted to a positive integer in the range from 0 to 255.

Parameter

- index—the position in the buffer.

Returns the value at the given position, or −1 if the position is invalid (beyond the end of the buffer).

```
public final byte[] getBytes()
```

Returns a copy of the byte array holding the buffered APDU. The byte array returned gets allocated the exact size of the buffered APDU. In contrast to getBuffer, manipulations in this array, of course, do not affect the internal buffer.

```
public final int getLength()
```

Gets the actual buffer length.

```
public final void setByte (int index,
                           int value)
```

Sets the byte at the specified position in the buffer. The byte is passed as an integer, for consistence with getByte. This method can only be used to modify an APDU already stored. It is not possible to set bytes beyond the end of the current APDU. The method will behave as a no-operation if this happens.

Parameters

- index—the position in the buffer;
- value—the byte to store there.

```
public final void setLength(int length)
    throws java.lang.IndexOutOfBoundsException
```

Sets the actual length the APDU buffer. This method can be used to cut off the end of the APDU. It can also be used to increase the size of the APDU when the caller has extended the APDU with something useful.

```
public java.lang.String toString()
```

Returns a humanly readable string representation of this APDU object.

B.2.2 Class CardID

```
public class CardID
extends java.lang.Object
```

Represents a smart card's ATR (answer to reset). In addition to the ATR itself, the slot where the card is inserted can be stored.

Fields

```
protected byte[] atr
```

The ATR represented.

```
protected byte[] historicals
```

The historical characters of the ATR. There can be at most 15 historical characters. If there are no historical characters, this attribute holds NULL, not an empty array.

```
protected int slotNr
```

A number representing the slot that holds the card with this ATR.

```
protected CardTerminal terminal
```

Points to the CardTerminal object associated with the slot.

```
protected java.lang.String cachedResult
```

The cached result of toString.

Constructors

```
public CardID(byte[] answerToResetResponse)
    throws CardTerminalException
```

Instantiates a new `cardID` representing the given ATR.

```
public CardID(CardTerminal terminal,
              int slotID,
              byte[] answerToResetResponse)
        throws CardTerminalException
```

As above, but with additional parameters.

Parameters

- `CardTerminal`—the terminal where the card with this ATR is inserted;
- `slotID`—the slot where the card with this ATR is inserted;
- `answerToResetResponse`—a byte array holding the ATR.

Methods

```
public byte[] getATR()
```

Returns a byte array that holds a copy of the ATR.

```
public byte[] getHistoricals()
```

Gets the byte array copy of the historical characters. Because they do not have to be printable characters, they are returned as a byte array, not as a string. NULL can be returned if there are none in the ATR.

```
public int getSlotID()
```

Gets the instantiating slot ID.

```
public boolean equals(java.lang.Object obj)
```

Compares this with another `CardID` object. True if the represented ATRs are equal.

```
public java.lang.String toString()
```

Returns a String representation of this `CardID` object. The string representation is cached in the probate field.

B.2.3 Class `CardTerminalFactory`

```
public abstract interface CardTerminalFactory
```

A `CardTerminalFactory` produces `CardTerminal` objects of a certain type attached to an address. The terminal manufacturer should provide his own version of a `CardTerminalFactory` that can produce appropriate `CardTerminal` objects. Each line in `opencard.properties` consists of three fields that are important to card terminal construction.

Fields

```
public final static int TERMINAL_NAME_ENTRY
```

First element in the terminal configuration array. Free-form string.

```
public final static int TERMINAL_TYPE_ENTRY
```

Second element in the terminal configuration array. Identifies which card terminal class has to be constructed.

```
public final static int TERMINAL_ADDRESS_ENTRY
```

Third element in the terminal configuration array. Represents some port where the card can be accessed (COM1, COM2, etc.).

Methods

```
public void
    createCardTerminals(CardTerminalRegistry ctr,
                        java.lang.String[] terminalInfo)
    throws CardTerminalException,
        TerminalInitException
```

Creates a specific `CardTerminal` object that knows how to handle a specific card terminal and register it with the `CardTerminalRegistry`.

Parameters

- ctr—the CardTerminalRegistry;
- terminalInfo—the parameter array for the terminal. The obligatory three parameters.

```
public void open()
        throws CardTerminalException
```

Initializes the CardTerminalFactory.

```
public void close()
        throws CardTerminalException
```

Closes the CardTerminalFactory.

B.2.4 Class CardTerminalRegistry

```
public final class CardTerminalRegistry
extends java.lang.Object
```

The CardTerminalRegistry keeps track of the installed card terminals registered within a system.

With the add() and remove() methods you can dynamically add and remove card terminals to and from the registry. Usually, however, this will be done automatically by SmartCard.start () on the basis of the information provided by the opencard.properties.

Field

```
protected java.util.Hashtable ctListeners
```

Table with all registered CTListener objects. A CTListener object represents something that waits for a card.

Methods

```
public void setObserver(Observer o)
```

The observer is the bridge to the opencard.core.event package, which creates events for card insertion and removal.

```
public static CardTerminalRegistry getRegistry()
```

Gets a reference to the system-wide `CardTerminalRegistry` object.

```
public void add(CardTerminal terminal)
    throws CardTerminalException
```

Adds a `CardTerminal` instance to the registry. Should be called by the `CardTerminalFactory` when creating a card terminal.

```
public void addPollable(Pollable p)
```

Adds a pollable card terminal to the observer's list of pollable terminals. Should be called by pollable terminals in their open() method.

```
public CardTerminal cardTerminalForName
(java.lang.String name)
```

Iterates through the registered terminals and searches for one with the given name. Returns NULL if none can be found.

```
public int countCardTerminals()
```

Gets the number of registered `CardTerminals`.

```
public java.util.Enumeration getCardTerminals()
```

Gets all registered `CardTerminal` instances.

```
public boolean remove(java.lang.String name)
    throws CardTerminalException
```

Removes the card terminal called name.

```
public boolean remove(CardTerminal terminal)
    throws CardTerminalException
```

Closes the card terminal and removes it from the registry.

```
public boolean removePollable(Pollable p)
```

Removes a pollable card terminal from the observer's list of terminals to be polled.

```
protected void cardInserted(CardTerminal terminal,
                              int slotID)
```

Notifies listeners that a card was inserted into a slot of a terminal.

Parameters

- `terminal`—the terminal where a card was inserted;
- `slot`—the slot where a card was inserted.

```
protected void cardRemoved(CardTerminal terminal,
                             int slotID)
```

Notifies listeners that a card was removed from a slot of this terminal (utility method).

B.2.5 Class `CardTerminal`

```
public abstract class CardTerminal
extends java.lang.Object
```

The `CardTerminal` class represents a physical card terminal. We assume that a certain card terminal provides at least one slot for a smart card or a transceiver for a contactless smart card.

`CardTerminal` objects are created by a `CardTerminalFactory` and registered in the `CardTerminalRegistry`. To get an enumeration of the available `CardTerminals`, invoke `CardTerminalRegistry.get-CardTerminals`.

Fields

```
protected final java.lang.String name
```

The free-form name of this card terminal (see `CardTerminalRegistry`). First identifier in a terminal line in `opencard.properties`.

```
protected final java.lang.String type
```

The type `string` of this terminal (second identifier in `opencard.properties`)

```
protected final java.lang.String address
```

Represents the string identifying the port to which the card reader is attached (third identifier).

Constructor

```
protected CardTerminal(java.lang.String name,
                       java.lang.String type,
                       java.lang.String address)
```

Instantiates a CardTerminal object and sets the usual three values.

Methods

```
public final java.util.Properties features()
```

Queries the card terminal about its features (name, type, address, number of slots, and additional terminal-specific features). Each feature is represented by a property.

```
public java.lang.String getAddress()
```

Returns the address of this CardTerminal.

```
public abstract CardID getCardID(int slotID)
    throws CardTerminalException
```

Returns the ATR response of the card inserted in slot designated slotID as a CardID. Before calling this method the caller should make sure that a card is present, otherwise NULL may be returned. This call should not block if no card is present.

```
public java.lang.String getName()
```

Returns the name associated with this CardTerminal.

```
public java.lang.String getType()
```

Returns the type of this CardTerminal.

```
public int getSlots()
```

Returns the number of slots belonging to this CardTerminal object.

```
public abstract boolean isCardPresent(int slotID)
    throws CardTerminalException
```

Checks whether there is a smart card present in a particular slot.

```
public boolean isSlotChannelAvailable(int slotID)
```

Checks whether a SlotChannel is available for a particular slot.

```
public abstract void open()
    throws CardTerminalException
```

Initializes the CardTerminal. Implementations of this method must carry out all steps required to set the concrete terminal into a proper state. After invoking this method it should be possible to communicate with the terminal properly.

```
public abstract void close()
    throws CardTerminalException
```

Closes the CardTerminal. Implementations of this method must carry out all steps required to close the concrete terminal and free resources held by it.

```
public final SlotChannel openSlotChannel(int slotID)
    throws InvalidSlotChannelException,
        java.lang.IndexOutOfBoundsException,
        CardTerminalException
```

Opens a SlotChannel on slot number slotID.

```
public final SlotChannel
    openSlotChannel(int slotID,
                    java.lang.Object lockHandle)
        throws InvalidSlotChannelException
            java.lang.IndexOutOfBoundsException,
            CardTerminalException
```

Opens a SlotChannel on slot number slotID using a lock handle.

```
public final void closeSlotChannel(SlotChannel sc)
    throws InvalidSlotChannelException,
        CardTerminalException
```

Closes a SlotChannel.

```
public final CardID reset(SlotChannel sc)
    throws InvalidSlotChannelException,
           CardTerminalException
```

Resets a smart card inserted in a slot. Calls `internalReset`, which actually does the reset.

```
public final ResponseAPDU sendAPDU(SlotChannel sc,
                                   CommandAPDU capdu)
    throws InvalidSlotChannelException,
           CardTerminalException
```

Sends a `CommandAPDU` to a slot and gets back the response.

```
public java.lang.String toString()
```

Returns a printable representation of this `CardTerminal` object.

```
protected void addSlots(int numberOfSlots)
throws CardTerminalException
```

Add slots to the `CardTerminal`. Used by the concrete `CardTerminal` implementations.

```
protected void cardRemoved(int slotID)
```

Notifies listeners that a card was removed from a slot of this terminal (utility method).

```
protected void cardInserted(int slotID)
```

Notifies listeners that a card was inserted into a slot of this terminal (utility method).

```
protected java.util.Properties
    internalFeatures(java.util.Properties features)
```

The `CardTerminal` internal `features()` method to be provided by the concrete implementation. This default implementation simply returns the parameter provided. Concrete implementations should override this method. Returns the enriched properties object.

```
protected void internalOpenSlotChannel(int slotID)
    throws CardTerminalException
```

The internal openSlotChannel method. The method internal-
OpenSlotChannel is executed at the beginning of openSlotChannel.

```
protected void
    internalOpenSlotChannel(int slotID,
                            java.lang.Object
                            lockHandle)
        throws CardTerminalException
```

The internal openSlotChannel method for locked terminals (default
implementation). Lockable terminals must overwrite this method. The
method internalOpenSlotChannel is executed at the beginning of
openSlotChannel.

```
void internalCloseSlotChannel(SlotChannel sc)
    throws CardTerminalException
```

The internal closeSlotChannel method. The method internal-
CloseSlotChannel is executed at the end of closeSlotChannel.

```
protected abstract CardID internalReset(int slot,
                                        int ms)
    throws CardTerminalException
```

The internal reset method to be provided by the concrete implementation.
In this version the ms parameter is ignored.

```
protected abstract ResponseAPDU
    internalSendAPDU(int slot,
                     CommandAPDU capdu,
                     int ms)
        throws CardTerminalException
```

The internalSendAPDU method to be provided by the concrete implementation. As usual the ms parameter can be ignored.

```
public ResponseAPDU
    sendVerifiedCommandAPDU(SlotChannel chann,
                            CommandAPDU capdu,
                            CHVControl vc,
```

```
                              int ms)
            throws CardTerminalException
```

Default implementation for deprecated method in interface `VerifiedAP-DUInterface`.

Note: An exception is thrown if this terminal does not implement `VerifiedAPDUInterface`.

B.2.6 Class SlotChannel

```
public final class SlotChannel
extends java.lang.Object
```

A `SlotChannel` serves a dual purpose: It is a gate object providing access to the smart card; in addition, it is used to send and receive APDUs and generally interact with the card.

Constructor

```
public SlotChannel(CardTerminal terminal,
                   int slotID,
                   java.lang.Object lockHandle)
                   Instantiates a SlotChannel.
```

Parameters

- `terminal`—the terminal to which the slot belongs;
- `slotID`—the number of the slot to which this `SlotChannel` is attached;
- `lockHandle`—the owner who holds a lock on the slot.

Methods

```
public java.lang.Object getScheduler()
```

Normally only used internally.

```
public void setScheduler(java.lang.Object scheduler)
```

Normally only used internally.

```
public ResponseAPDU sendAPDU(CommandAPDU capdu)
throws CardTerminalException
```

Sends a CommandAPDU to this SlotChannel.

```
public int getSlotNumber()
```

Returns the slot number of the associated slot.

```
public CardTerminal getCardTerminal()
```

Returns the CardTerminal.

```
public java.lang.Object getLockHandle()
```

Returns the slot owner.

```
public CardID reset()
throws CardTerminalException
```

Resets the smart card attached to this SlotChannel's slot.

```
public boolean isOpen()
```

Checks whether this SlotChannel is open.

```
public void close()
    throws CardTerminalException
```

Closes this SlotChannel. Once a SlotChannel is closed, it cannot be used again.

```
public CardID getCardID()
    throws CardTerminalException
```

Returns the CardID object of the inserted smart card.

```
public java.lang.String toString()
```

Returns a string representation of this instance.

B.3 Package `opencard.opt.service`

B.3.1 Class `OCF11CardServiceFactory`

This class is used for backward compatibility to OCF 1.1.x to instantiate "old-fashioned card services." It is mentioned here only for historical reasons.

```
public abstract class OCF11CardServiceFactory
extends opencard.core.service.CardServiceFactory
```

Constructor

```
public OCF11CardServiceFactory()
```

Instantiates a `CardServiceFactory`.

Methods (only the old ones)

```
protected abstract boolean
    knows(opencard.core.terminal.CardID cid)
```

Returns a simple yes/no decision if the card service can be identified by looking at the ATR provided by the cid.

```
protected abstract java.util.Enumeration
    cardServiceClasses(opencard.core.terminal.CardID cid)
```

Returns an enumeration of known `CardService` classes.

B.4 Package `opencard.opt.util`

B.4.1 Class `TLV`

```
public class TLV
extends java.lang.Object
```

This class represents a TLV structure. There are methods for creating trees consisting of TLV objects from ASN.1 BER-encoded byte sequences and for creating byte sequences from TLV object trees. All manipulations are done on the tree structure.

Constructors

```
public TLV()
```

Creates an empty TLV object.

```
public TLV(byte[] binary)
```

Creates a TLV object from an ASN.1 BER-encoded byte array (defined in ISO 8825).

```
public TLV(Tag tag,
           byte[] value)
```

Creates a TLV object from the given tag object and data. If the given tag object has the constructed bit set, the result will be a TLV tree, otherwise it is just a primitive TLV that contains the data given in the value field.

```
public TLV(Tag tag,
           int number)
```

Creates a primitive TLV object from a given tag and positive integer.

```
public TLV(Tag tag,
           TLV tlv)
```

Creates a constructed TLV object from the given tag object and TLV object to be contained.

Methods

```
public TLV add(TLV tlv)
```

Adds the given TLV object to this TLV instance (only if constructed).

```
public TLV findTag(Tag tag,
                   TLV cursor)
```

Searches for a given tag value and returns the first TLV found.

Parameters

- tag—the tag object representing the tag to be searched for, NULL for any tag;

- cursor—a reference to a TLV object where the search should start; if NULL, the search is started with the child of this TLV instance.

Returns the first TLV object found, which has the given tag value; NULL if no match is found.

```
public static void fromBinary(byte[] binary,
                              int[] offset,
                              TLV tlv,
                              TLV parent)
```

Reads a TLV object from a binary representation.

Parameters

- binary—a byte array containing the binary representation of a TLV structure, encoded conforming to the ASN.1 Basic Encoding Rules defined in ISO 8825;

- offset—an integer value giving the offset where the binary representation starts;

- tlv—the TLV object to be read from the binary representation;

- parent—the TLV object representing the parent of the object to be read.

Returns an integer value giving the offset of the end of the binary representation read into the TLV object.

```
public static int lenBytes(int length)
```

Returns the number of bytes required for coding the passed integer value as described in the ASN.1 Basic Encoding Rules.

```
public int length()
```

Gets the length of this TLV's value field in bytes.

```
public static byte[] lengthToBinary(int length)
```

BER codes the length of this TLV object.

Parameters

- binary—the byte array to which the BER-coded length field shall be added;
- offset—the offset, where the BER-coded length field shall be added.

```
public void setValue(byte[] newValue)
```

Sets the value field of this TLV from the byte array.

```
public Tag tag()
```

Gets the tag of this TLV.

```
public byte[] toBinary()
```

Returns a byte array containing the BER-coded representation of this TLV instance.

```
public byte[] toBinaryContent()
```

Returns a byte array containing the BER-coded representation of the value field of this TLV instance.

```
public java.lang.String toString()
```

Returns a string representation of this TLV object.

```
public byte[] valueAsByteArray()
```

Gets the value field of this TLV as a byte array.

```
public int valueAsNumber()
```

Gets the value of this TLV as a positive integer number.

B.4.2 Class Tag

```
public class Tag
extends java.lang.Object
```

This class represents tags as defined in the Basic Encoding Rules for ASN.1 defined in ISO 8825. A `tag` consists of two class bits (0–4),[1] and a flag indicating whether the TLV is constructed or primitive.

The encoding is like this (C = class bit, c = composed flag, X = data bit):

- Range from 0x0–0x1E:

```
C C c X X X X X
```

- Range from 0x1F–0x80:

```
C C c 1 1 1 1 1 0 X X X X X X X
```

- Range from 0x81–0x4000:

```
C C c 1 X X X X 1 X X X X X X X 0 X X X X X X X
```

- Range from 0x4001–0x200000:

```
C C c 1 X X X X 1 X X X X X X X 1 X X X X X X X 0 X X
X X X X X
```

- Range from 0x200001–0x10000000:

```
C C c 1 X X X X 1 X X X X X X X 1 X X X X X X X 0 X X
X X X X X 0 X X X X X X X
```

Constructors

```
public Tag()
```

Creates a NULL tag.

```
public Tag(Tag t)
```

Clones a tag.

---

1. I don't know how to encode 5 values in 2 bits, but so says the OCF documentation.

```
public Tag(int tag,
            byte tagClass,
            boolean constructed)
```

Creates a tag from a given `tag` value, class, and constructed flag.

```
public Tag(byte[] binary)
```

Creates a tag from binary representation.

```
public Tag(byte[] binary,
            int[] offset)
```

As above, but with an offset pointing to where to start.

Methods

```
public int size()
```

Returns the number of bytes required to BER code the `tag` value.

```
public void fromBinary(byte[] binary,
                        int[] offset)
```

Initializes the `tag` object from a BER-coded binary representation.

Parameters

- `binary`—a byte array containing the BER-coded tag;

- `offset`—an integer giving an offset into the byte array from where to start.

```
public byte[] getBytes()
```

Gets a byte array representing the tag.

```
public void toBinary(byte[] binary,
                      int[] offset)
```

Converts the tag to binary representation.

```
public void set(int tag,
                byte tagclass,
                boolean constructed)
```

Sets the tag number, class, and constructed flag of this tag to the given values.

Parameters

- tag—an integer value giving the tag value;
- tagclass—a byte value giving the class;
- constructed—a Boolean value representing the constructed flag.

```
public void setConstructed(boolean constructed)
```

Sets the constructed flag of this tag to the given value.

```
public int code()
```

Gets the code of the tag.

```
public boolean isConstructed()
```

Checks whether this tag is constructed.

```
public int hashCode()
```

Computes a hash code for this tag.

```
public boolean equals(java.lang.Object o)
```

Checks for equality.

```
public java.lang.String toString()
```

Gets a string representation for this tag.

Reference

[1]   OCF Version 1.2 Apidocs; available at http://www.opencard.org/docs/1.2/index.html.

List of Acronyms

AAC application authentication cryptogram

ABS acrylonitrile-butadiene-styrene

AC application cryptogram

ADF application definition file

AEF application elementary file

AES advanced encryption standard

AFL application file locator

AID application identifier or applet identifier

AIP application interchange profile

APDU application protocol data unit

API application programming interface

ARQC authorization request cryptogram

ASCII   American Standard Code for Information Interchange

ASN.1   Abstract Syntax Notation One

ATC   application transaction counter

ATM   automatic teller machine

ATR   answer to reset

AUC   application usage control

AVN   application version number

AWT   abstract window toolkit

BER   basic encoding rules

CA   certificate authority

CAD   card accepting device

CAP   converted applet

CBC   cipher block chaining

CC   common criteria

CDOL   card risk management data object list

CEN   European Committee for Standardization

CEPS   common electronic purse specifications

CHV   cardholder verification

CIAL   card issuer action code

CID   cryptogram information data

CLA class byte

CMS card management system

CPU central processing unit

CT-API card terminal application programming interface

CTCPEC *Canadian Trusted Computer Product Evaluation Criteria*

CVM cardholder verification method

CVR card verification results

DAC data authentication code

DDA dynamic data authentication

DDOL dynamic data object list

DDF directory definition file

DES data encryption standard

DF dedicated file

DFA differential fault analysis

DOL data object list

DPA differential power analysis

DSA digital signature algorithm

EAL evaluation assurance levels

ECB electronic code book

EDC error detection code

EDE   encrypt-decrypt-encrypt

EEPROM   electrically erasable programmable read-only memory

EESSI   European Electronic Signature Standardisation Initiative

EF   elementary file

EMV   Europay/MasterCard/Visa

ETSI   European Telecommunications Standards Institute

etu   elementary time unit

EU   European Union

EUROSMART   European Smart Card Industry Initiative

FC   federal criteria

FCI   file control information

FID   file identifier

GPRS   general packet radio service

GSM   Global System for Mobile Communications

GUI   graphical user interface

I/O   input/output

IAC   issuer action codes

IAD   issuer application data

IAIK   Institut für Angewandte Informationsverarbeitung und Kommuni-kationstechnologie (German)

IC   integrated circuit

ICC integrated circuit card

IDEA international data encryption algorithm

IEC International Electrotechnical Commission

IEEE Institute of Electrical and Electronics Engineers

IEP intersector electronic purse

IFD interface device

INS instruction byte

ISO International Organization for Standardization

ISSS Information Society Standardization System

ITSEC Information Technology Security Evaluation Criteria

JAR Java archive

JC Java card

JCA Java Card Forum

JCM Java card management

JCRE Java card run-time environment

JCVM Java card virtual machine

JIT just in time

JNI Java Native Interface

JVM Java virtual machine

LC length of command

LCD   liquid crystal display

LE   expected length of response

LEN   length byte

MAC   message authentication code

MEL   Multos execution language

MF   master file

MPEG   Moving Pictures Experts Group

NAD   node address byte

OCF   OpenCard Framework

OS   operating system

OTA   open terminal architecture

PAN   primary account number

PAS   publicly available specification

PC   polycarbonate

PC/SC   personal computer/smart card

PCB   protocol control byte

PCMCIA   Personal Computer Memory Card International Association

PCT   personal chipcard terminal

PD   purchase device

PDOL   processing options data object list

PET   polyethylene-terephthalate

PGP   pretty good privacy

PIN   personal identification number

PIX   proprietary application identifier extension

PKA   public key algorithm

PKCS   public key cryptography standard

PMA   platform management architecture

PODL   processing options data object list

POS   point of sale

PP   protection profile

PPS   protocol and parameters selection

PSE   payment systems environment

PTS   protocol type selection

PVC   polyvinylchloride

RCP   reference control parameter

RAM   random access memory

RC4   Ron's Code 4

RFU   reserved for future use

RID   registered identifier

ROM   read-only memory

RSA   Rivest, Shamir, Adleman

SCD   signature-creation data

SCUGPP   Smart Card Security User Group's Protection Profile

SDA   static data authentication

SET   secure electronic transactions

SFI   short file identifier

SIM   subscriber identity module

SIO   shareable interface object

SM   security module

SMS   short message service

SOF   strength of functions

SPA   simple power analysis

SSCD-PP   secure signature-creation device protection profile

SSL   secure sockets layer

ST   security target

STIP   small terminal interoperability platform

SVD   signature-verification data

SW   status word

TAN   transaction authentication numbers

TC   transaction certificate

TCSEC   Trusted Computer System Evaluation Criteria

TLV   tag-length-value

TOCF   Tiny OpenCard Framework

TOE   target of evaluation

TPDU   transport protocol data unit

TTP   trusted third party

UML   universal modeling language

USB   universal serial bus

USSD   unstructured supplementary services data

VIS   Visa ICC specification

VOP   Visa open platform

VSCPP   Visa smart card protection profile

WfSC   Windows for Smart Cards

XOR   exclusive-OR

About the Authors

Vesna Hassler, Ph.D., teaches network security and electronic payment systems at Vienna University of Technology, Austria. She has published a number of papers on cryptography, network security, payment systems, and smart cards, and one book, *Security Fundamentals for E-Commerce* (Artech House, 2001). She has managed research projects in the areas of security and software engineering and worked as a consultant on the electronic signature card project led by major Austrian banks and the Austrian National Bank. Currently she holds a position at the Austrian Center for Secure Information Technology (A-SIT) where she is involved in security evaluation of secure signature creation devices (e.g., smart cards for electronic signatures) and standardization activities in this area.

Martin Manninger, Ph.D., holds M.Sc. degrees in the fields of electrical engineering, economics, and information technology. In his Ph.D. thesis he focused on Internet payment and smart card technology, which were his main research interests during his time as a research assistant at Vienna University of Technology, Austria. Currently he manages smart card projects at Austria Card, Ltd., the leading smart card manufacturer in Austria, where his areas of responsibility include bank cards, digital signature cards, and contactless cards. He has published several papers on field bus systems, Internet payment systems, and smart cards, and coauthored one book, *Electronic Commerce—Die Technik. Technologie, Design und Implementierung* (with M. Göschka, C. Schwaiger, and D. Dietrich; Hüthig Verlag, 2000, in German).

Mikhail Gordeev, Ph.D., holds an M.Sc. degree in computer science. He received his Ph.D. from Vienna University of Technology, Austria. His Ph.D. thesis was dedicated to communication security issues in heterogeneous network environments. Currently working as an independent consultant in the areas of network security and smart card technology, he is engaged in a number of industrial R&D projects developing smart card–based security solutions and secure networking concepts. He has published a number of papers and given talks at international conferences on those topics.

Christoph Müller, M.Sc., earned his master's degree in computer science at Vienna University of Technology, Austria. His thesis concerned developing a system to make Quick (Austrian electronic purse) transactions available over the Internet (Telequick) in a secure way. After several years of developing and testing in the electronic signature card project led by major Austrian banks and the Austrian National Bank, he is currently working for Xsoft GmbH, a leading Austrian company, in IT security as well as computer and network security.

Pedrick Moore, the technical editor for this book, is a freelance language consultant with more than 20 years of experience in coaching nonnative speakers of English. A graduate of the University of Virginia (B.A., German), she edits and translates business publications, research and academic papers, and film scripts. Her clients include corporations, research institutions, academics, artists, and professionals. She divides her time between Vienna and a vineyard in southern England.

Index

Recent Titles in the Artech House Computing Library

Software Verification and Validation for Practitioners and Managers, Second Edition, Steven R. Rakitin

Strategic Software Production with Domain-Oriented Reuse, Paolo Predonzani, Giancarlo Succi, and Tullio Vernazza

Systems Modeling for Business Process Improvement, David Bustard, Peter Kawalek, and Mark Norris, editors

User-Centered Information Design for Improved Software Usability, Pradeep Henry

Workflow Modeling: Tools for Process Improvement and Application Development, Alec Sharp and Patrick McDermott

For further information on these and other Artech House titles, including previously considered out-of-print books now available through our In-Print-Forever® (IPF®) program, contact:

Artech House	Artech House
685 Canton Street	46 Gillingham Street
Norwood, MA 02062	London SW1V 1AH UK
Phone: 781-769-9750	Phone: +44 (0)20 7596-8750
Fax: 781-769-6334	Fax: +44 (0)20 7630-0166
e-mail: artech@artechhouse.com	e-mail: artech-uk@artechhouse.com

Find us on the World Wide Web at:
www.artechhouse.com